Essentials of Plant Cell Biology

Essentials of Plant Cell Biology

William Cole

STATES
ACADEMIC PRESS
www.statesacademicpress.com

States Academic Press,
109 South 5th Street,
Brooklyn, NY 11249, USA

Visit us on the World Wide Web at:
www.statesacademicpress.com

ISBN: 978-1-63989-188-7 (Hardback)

Cataloging-in-Publication Data

Essentials of plant cell biology / William Cole.
p. cm.
Includes bibliographical references and index.
ISBN 978-1-63989-188-7
1. Plants--Cytology. 2. Plant cells and tissues. 3. Plant anatomy.
4. Cells. I. Cole, William.
QK725 .E87 2022
581.072 4--dc23

Table of Contents

Preface

Plant cell biology is a natural science which studies the physical structure as well as the chemical and physiological processes related to plants at the cellular level. Plant cells are eukaryotic cells which are found in the photosynthetic eukaryotes of the Plantae kingdom. Some of their distinctive features are the presence of a cellulosic cell wall, capability to store starch, presence of plastids and ability to perform photosynthesis. Plant cells and tissues can be majorly classified into parenchyma cells, collenchyma cells, sclerenchyma cells, xylem, phloem and epidermis. This book presents this complex subject in the most comprehensible and easy to understand language. The book aims to shed light on some of the fundamental concepts of plant cell biology. It will serve as a valuable source of information for those interested in this field.

A foreword of all Chapters of the book is provided below:

Chapter 1 - Plant cell biology is concerned with the study of the structure, growth, signaling and differentiation of plant cells. Some of the areas of study related to plants cells are plant tissues, cell theory, cell cycle, photosynthesis and plant hormones. This is an introductory chapter which will briefly introduce all these significant aspects of plant cell biology; **Chapter 2** - Plants cells are eukaryotic cells that have cell walls composed of cellulose, hemicellulose and pectin. Specific plant cells perform specific function in plants. For example, meristemic cells are totipotent and perform continued cell division, and collenchyma cells provide support and structure. This chapter covers different types of cells such as meristematic cells, parenchyma cells, collenchyma cells, sclerenchyma cells, epidermal cells, xylem cells and phloem cells; **Chapter 3** - Nucleus is the membrane-bound organelle which encloses the genetic material of eukaryotic organisms. Eukaryotic cells generally have a single nucleus but some cells such as red blood cells lack nucleus and some cells, for example, osteoclasts have multiple nuclei. This chapter has been carefully written to provide an easy understanding of the varied facets of nucleus such as nuclear envelope and nuclear pore complex; **Chapter 4** - Plant cell wall is a structure present around the plasma membrane of plant cells. It is a rigid structure and provides tensile strength to the cell, and protects against mechanical and osmotic stress. Plant cell walls are mainly composed of cellulose, which is a linear polymer of glucose. This chapter discusses in detail the concepts related to plant cell wall; **Chapter 5** - Plastids are the membrane bound organelles which are mainly responsible for the manufacturing and storage of food in plants. They also contain pigments used in photosynthesis. The types of pigments of the plastid are responsible for determining the cell color. All the diverse principles of plastids have been carefully analyzed in this chapter; **Chapter 6** - The process of transmitting a chemical or physical signal through a series of molecular events which result in a cellular response is termed as signal transduction. The various facets related to signal transduction in plants such as plant receptors, signaling pathways and secondary messengers are discussed in this chapter in an easy-to-understand manner.

I would like to thank the entire editorial team who made sincere efforts for this book and my family who supported me in my efforts of working on this book. I take this opportunity to thank all those who have been a guiding force throughout my life.

William Cole

Chapter 1

Plant Cell Biology: An Introduction

Plant cell biology is concerned with the study of the structure, growth, signaling and differentiation of plant cells. Some of the areas of study related to plants cells are plant tissues, cell theory, cell cycle, photosynthesis and plant hormones. This is an introductory chapter which will briefly introduce all these significant aspects of plant cell biology.

Plant cell biology is the study of all type and aspects of plant cells. Plant Cells are the basic building block of plant life, and they carry out all of the essential functions necessary for plant survival. It is particularly concerned with structure, growth, signalling, differentiations and death of plant cells.

Plant cells, like animal cells, are eukaryotic, meaning they have a membrane-bound nucleus and organelles. The following is a brief survey of some of the major characteristics of plant cells.

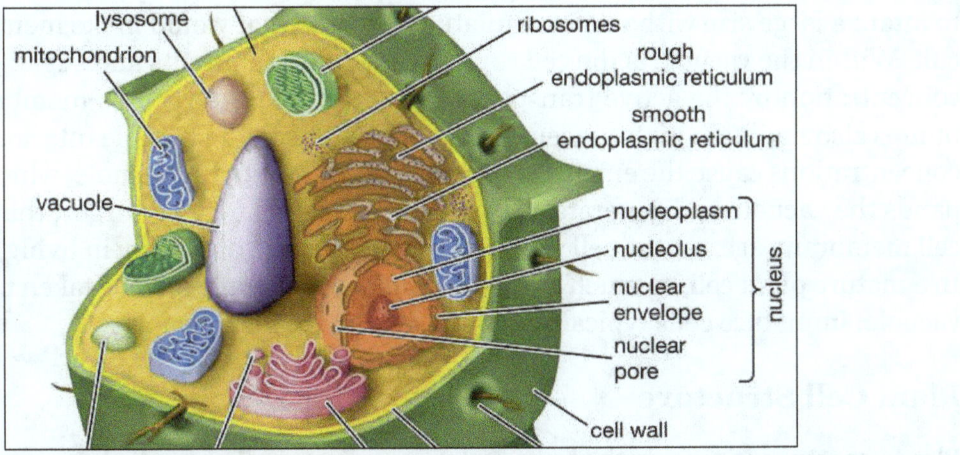

Plant cell: Cutaway drawing of a plant cell, showing the cell wall and internal organelles.

Unlike animal cells, plant cells have a cell wall surrounding the cell membrane. Although often perceived as an inactive product serving mainly mechanical and structural purposes, the cell wall actually has a multitude of functions upon which plant life depends. Plant cell walls are composed of cellulose, which sets them apart from other organisms with cell walls, such as bacteria (peptidoglycan) and fungi (chitin). Algal cell walls are similar to those of plants, and many contain specific polysaccharides that are useful for taxonomy.

Plant cells can be distinguished from most other cells by the presence of chloroplasts, which are also found in certain algae. A chloroplast is a type of plastid (a saclike organelle with a double membrane) that serves as the site of photosynthesis, the process by which energy from the Sun is converted into chemical energy for growth. Chloroplasts contain the pigment chlorophyll to absorb light energy. In plants, these essential organelles occur in all green tissues, though they are concentrated particularly in the parenchyma cells of leaves.

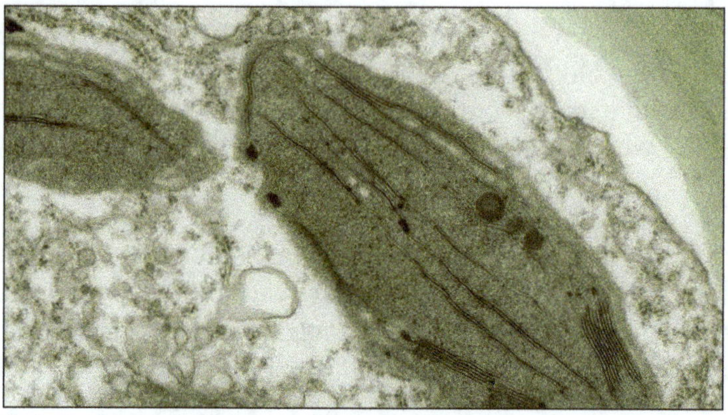

Chloroplast: False-colour transmission electron micrograph of a chloroplast in a bean leaf.

Another important characteristic of many plant cells is the presence of one or more large vacuoles. Vacuoles are storage organelles, and those in plant cells enable them to attain a large size without accumulating the bulk that would make metabolism difficult. Within the vacuole is the cell sap, a water solution of salts and sugars kept at high concentration by the active transport of ions through the vacuole membrane. Proton pumps also maintain high concentrations of protons in the vacuole interior. These high concentrations cause the entry, via osmosis, of water into the vacuole, which in turn expands the vacuole and generates a hydrostatic pressure, called turgor, that presses the cell membrane against the cell wall. Turgor is the cause of rigidity in living plant tissue. In a mature plant cell, as much as 90 per cent of cell volume may be taken up by a single vacuole; immature cells typically contain several smaller vacuoles.

Plant Cell Structure

Plants are unique among the eukaryotes, organisms whose cells have membrane-enclosed nuclei and organelles, because they can manufacture their own food. Chlorophyll, which gives plants their green color, enables them to use sunlight to convert water and carbon dioxide into sugars and carbohydrates, chemicals the cell uses for fuel.

Like the fungi, another kingdom of eukaryotes, plant cells have retained the protective cell wall structure of their prokaryotic ancestors. The basic plant cell shares a similar construction motif with the typical eukaryote cell, but does not have centrioles, lysosomes, intermediate filaments, cilia, or flagella, as does the animal cell. Plant cells do, however, have a number of other specialized structures, including a rigid cell wall, central vacuole, plasmodesmata, and chloroplasts. Although plants (and their typical cells) are non-motile, some species produce gametes that do exhibit flagella and are, therefore, able to move about.

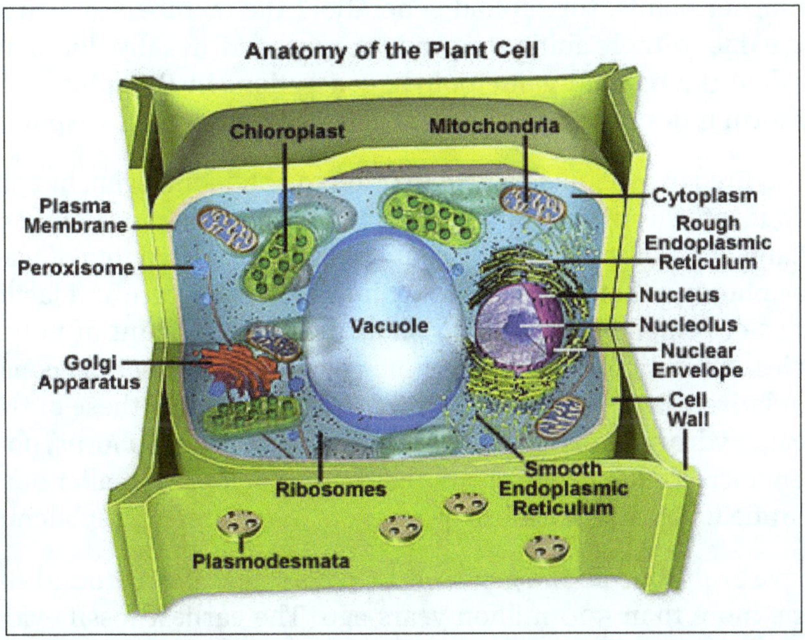

Plants can be broadly categorized into two basic types: Vascular and nonvascular. Vascular plants are considered to be more advanced than nonvascular plants because they have evolved specialized tissues, namely xylem, which is involved in structural support and water conduction, and phloem, which functions in food conduction. Consequently, they also possess roots, stems, and leaves, representing a higher form of organization that is characteristically absent in plants lacking vascular tissues. The nonvascular plants, members of the division Bryophyta, are usually no more than an inch or two in height because they do not have adequate support, which is provided by vascular tissues to other plants, to grow bigger. They also are more dependent on the environment that surrounds them to maintain appropriate amounts of moisture and, therefore, tend to inhabit damp, shady areas.

It is estimated that there are at least 260,000 species of plants in the world today. They range in size and complexity from small, nonvascular mosses to giant sequoia trees, the largest living organisms, growing as tall as 330 feet (100 meters). Only a tiny percentage of those species are directly used by people for food, shelter, fibre, and medicine. Nonetheless, plants are the basis for the Earth's ecosystem and food web, and without them complex animal life forms (such as humans) could never have evolved. Indeed, all living organisms are dependent either directly or indirectly on the energy produced by photosynthesis, and the by-product of this process, oxygen, is essential to animals. Plants also reduce the amount of carbon dioxide present in the atmosphere; hinder soil erosion, and influence water levels and quality.

Plants exhibit life cycles that involve alternating generations of diploid forms, which contain paired chromosome sets in their cell nuclei, and haploid forms, which only possess a single set. Generally these two forms of a plant are very dissimilar in appearance. In higher plants, the diploid generation, the members of which are known as sporophytes due to their ability to produce spores, is usually dominant and more recognizable than the haploid gametophyte generation. In Bryophytes, however, the gametophyte form is dominant and physiologically necessary to the sporophyte form.

Animals are required to consume protein in order to obtain nitrogen, but plants are able to utilize inorganic forms of the element and, therefore, do not need an outside source of protein. Plants do, however, usually require significant amounts of water, which is needed for the photosynthetic process, to maintain cell structure and facilitate growth, and as a means of bringing nutrients to plant cells. The amount of nutrients needed by plant species varies significantly, but nine elements are generally considered to be necessary in relatively large amounts. Termed macro elements, these nutrients include calcium, carbon, hydrogen, magnesium, nitrogen, oxygen, phosphorus, potassium, and sulphur. Seven microelements, which are required by plants in smaller quantities, have also been identified: boron, chlorine, copper, iron, manganese, molybdenum, and zinc.

Thought to have evolved from the green algae, plants have been around since the early Palaeozoic era, more than 500 million years ago. The earliest fossil evidence of land plants dates to the Ordovician Period (505 to 438 million years ago). By the Carboniferous Period, about 355 million years ago, most of the Earth was covered by forests of primitive vascular plants, such as lycopods (scale trees) and gymnosperms (pine trees, ginkgos). Angiosperms, the flowering plants, didn't develop until the end of the Cretaceous Period, about 65 million years ago—just as the dinosaurs became extinct:

- Cell Wall: Like their prokaryotic ancestors, plant cells have a rigid wall surrounding the plasma membrane. It is a far more complex structure, however, and serves a variety of functions, from protecting the cell to regulating the life cycle of the plant organism.
- Chloroplasts: The most important characteristic of plants is their ability to photosynthesize, in effect, to make their own food by converting light energy into chemical energy. This process is carried out in specialized organelles called chloroplasts.
- Endoplasmic Reticulum: The endoplasmic reticulum is a network of sacs that manufactures, processes, and transports chemical compounds for use inside and outside of the cell. It is connected to the double-layered nuclear envelope, providing a pipeline between the nucleus and the cytoplasm. In plants, the endoplasmic reticulum also connects between cells via the plasmodesmata.
- Golgi Apparatus: The Golgi apparatus is the distribution and shipping department for the cell's chemical products. It modifies proteins and fats built in the endoplasmic reticulum and prepares them for export as outside of the cell.

- Microfilaments: Microfilaments are solid rods made of globular proteins called actin. These filaments are primarily structural in function and are an important component of the cytoskeleton.
- Microtubules: These straight, hollow cylinders are found throughout the cytoplasm of all eukaryotic cells (prokaryotes don't have them) and carry out a variety of functions, ranging from transport to structural support.
- Mitochondria: Mitochondria are oblong shaped organelles found in the cytoplasm of all eukaryotic cells. In plant cells, they break down carbohydrate and sugar molecules to provide energy, particularly when light isn't available for the chloroplasts to produce energy.
- Nucleus: The nucleus is a highly specialized organelle that serves as the information processing and administrative center of the cell. This organelle has two major functions: It stores the cell's hereditary material, or DNA, and it coordinates the cell's activities, which include growth, intermediary metabolism, protein synthesis, and reproduction (cell division).
- Peroxisomes: Microbodies are a diverse group of organelles that are found in the cytoplasm, roughly spherical and bound by a single membrane. There are several types of microbodies but peroxisomes are the most common.
- Plasmodesmata: Plasmodesmata are small tubes that connect plant cells to each other, providing living bridges between cells.
- Plasma Membrane: All living cells have a plasma membrane that encloses their contents. In prokaryotes and plants, the membrane is the inner layer of protection surrounded by a rigid cell wall. These membranes also regulate the passage of molecules in and out of the cells.
- Ribosomes: All living cells contain ribosomes, tiny organelles composed of approximately 60 percent RNA and 40 percent protein. In eukaryotes, ribosomes are made of four strands of RNA. In prokaryotes, they consist of three strands of RNA.
- Vacuole: Each plant cell has a large, single vacuole that stores compounds, helps in plant growth, and plays an important structural role for the plant.

Leaf Tissue Organization

The plant body is divided into several organs: roots, stems, and leaves. The leaves are the primary photosynthetic organs of plants, serving as key sites where energy from light is converted into chemical energy. Similar to the other organs of a plant, a leaf is comprised of three basic tissue systems, including the dermal, vascular, and ground tissue systems. These three motifs are continuous throughout an entire plant, but their properties vary significantly based upon the organ type in which they are located.

Plant Tissues

Plant tissues are collections of similar cells performing related functions. Different plant tissues will have their own specialized roles and can be combined with other tissues to form organs such as flowers, fruit, stem, and leaves. Two major types of plant tissue include meristematic and permanent tissue.

Meristematic tissue, the primary growth tissue in plants, is capable of self-renewal and indefinite cell division. Every cell in the plant originates from a meristem. Meristematic tissue is classified into one of three types depending on its location inside the plant - apical, lateral, and intercalary. Apical meristems are meristematic tissue located at the tip of root and stem, which enable elongation of plant length. Lateral meristems are present in the radial portion of the stem and root and increase the thickness or girth of the maturing plant. Intercalary meristems occur only in monocots at the base of the internode and leaf blade. The intercalary meristems increase the length of the leaf blade.

Permanent plant tissues are either simple (composed of similar types of cells) or complex (consisting of different kinds of cells). For example, dermal tissue is a simple permanent tissue that forms the outer protective covering. It protects the plant from physical damage and enables gas exchange. In non-woody plants, the dermal tissue is a layer of tightly packed cells called the epidermis. The cuticle, a waxy epidermal coating, is present on leaves and stems that prevent water loss. The epidermis has unique functions in different plant organs. For example, roots, water, and minerals absorbed from the soil enter through the epidermis.

Vascular tissue, in contrast, is an example of complex tissue that enables the transport of water and minerals through the plant. The vascular system is composed of two specialized conducting vessels: xylem and phloem. Xylem conducts water and minerals from the roots to different parts of the plant, and itself consists of three types of cells: xylem vessel, tracheids (both of which hold water), and xylem parenchyma. Phloem conducts organic compounds from the site of photosynthesis to different parts of the plant. It includes four different types of cells: sieve cells (which conduct photosynthesis), phloem parenchyma, companion cells, and phloem fibers. In the stem, the xylem and phloem together form a structure called a vascular bundle. In roots, this is called the vascular cylinder or vascular stele.

Parenchyma, Collenchyma and Sclerenchyma

Plant anatomy divides the organism into four primary organs - root, stem, leaf, and flower. These can subsequently be divided into three tissue types. For example, leaves consist of three different tissues -dermal, vascular, and ground tissues. Further, these tissues are each composed of up to three cell types-parenchyma, sclerenchyma, or collenchyma.

Parenchyma cells are living, metabolically active, and usually bounded by a thin and flexible primary cell wall. In general, parenchyma cells account for 90 percent of the cells found in herbaceous seed plants. These often occur in the cortex or pith of stems or roots, and the fleshy tissue of many fruits. Most parenchyma cells retain the ability to divide, making them essential in wound healing and tissue regeneration. Moreover, parenchyma cells perform specialized functions in plants such as photosynthesis, storage, or transport, and aid the vascular tissue by forming a route to exchange nutrients within or between xylem and phloem.

Collenchyma cells are also living, and are elongated in structure, consisting of an irregular thick cell wall that provides support and structure to the plant. These are the least common cell type and have cell walls composed of cellulose and pectin. The epidermal tissue of young stem and leaf veins consists of collenchyma cells. There are three general classifications of collenchyma cells, depending on location and pattern of cell wall thickenings - angular, annular, lamellar, and lacunar.

Sclerenchyma cells form protective or supportive tissue in higher plants. At maturity, these cells have limited physiological activity and are usually dead. Sclerenchyma cells have a cell wall with a thickened secondary layer made up of cellulose, hemicellulose, and lignin. The orientation of the cellulose provides a diverse combination of strength, flexibility, and stiffness in plant organs subjected to different compressive and tensile forces. Sclerenchyma occurs in three different forms - fibers, sclereids, and water-conducting sclerenchyma.

Cell Theory

The modern cell theory, one of the fundamental generalizations of biology, holds that:

- All organisms are composed of one or more cells.
- New cells come from pre-existing cells; life forms today have descended in unbroken continuity from the first primitive cells that arose on earth more than 3.5 billion years ago.
- Hereditary information passes from parent cell to daughter cell.
- The fundamental biochemical reactions of life take place within cells.

Methods of Studying Cells

Microscopes: Cytology—the scientific study of cells—has progressed simultaneously with the development of better, more powerful microscopes and cell preparation techniques. Light microscopes, invented in the 1500s, today can magnify in the range of

100x to 1000x and are at the maximum of their resolving power at about 0.2 Φm (micrometer = $^{1}/_{1,000,000}$ meter). They are limited from further improvement by the wavelength of visible light that is used to create the image (the shorter the wavelength, the greater the resolution).

Beams of electrons have shorter wavelengths than visible light, hence electron microscopes (which focus streams of electrons) are able to achieve resolutions of 2–0.4 nm (nanometer = $^{1}/_{1,000,000,000}$ meter = 1/1000 Φm) and magnifications of over 100,000x. One type of electron microscope, the transmission electron microscope (TEM), focuses a beam of electrons *through* an object much as a light beam is used in light microscopes, whereas a scanning electron microscope (SEM) forms an image from electrons that are scattered from the *surface* of a specimen.

Laboratory techniques: In light microscopy dyes that stain particular chemicals are commonly used to identify cellular structures in thin sections of tissue mounted on microscope slides. Specimens observed by electron microscopy are cut in much thinner sections (electrons don't penetrate material well) and examined in a vacuum. These preparations cause problems of interpretation of the microscopic image. Does the material *really* look like this in living cells or has preparation altered its appearance? Living material cannot be examined using electron microscopes so the answer is moot.

Today's understanding of cells is based on two approaches:

- The sometimes very difficult biochemical analyses of cells.
- The increasing greater magnification of their constituent parts.

Cell Size

Plant cells vary in size from about 7 Φm—the diameter of dividing cells at the tips of roots and shoots—to fiber cells a few micrometers wide but a meter or more in length. Most plant (and animal) cells are between 10–100 Φm and thus too small to be seen without magnification since human eyes have a resolving power (resolution) of about 100 Φm. Resolving power is the ability to see adjacent objects as separate; magnification simply makes things larger. Two things closer together than 100 Φm will look like one to the human eye unless a microscope with a resolving power of less than 100 Φm is used.

Limits to Cell Growth

Plant cells are separated from their environment by a cell wall inside of which a semi-liquid plasma membrane surrounds the cytoplasm and the nucleus. As cells grow, their volume increases faster than their surface area, a geometric reality (volume increases as the cube of the diameter, surface as the square of the diameter in the surface: volume ratio). Since materials used and released in the metabolism of the cell must pass through its surface, a functional upper size limit to growth is reached; cells can grow no larger in volume than the membrane area can support. Distribute the same volume of cytoplasm contained in one large cell into several smaller cells and the surface area increases, the cell can function, and ultimately, multicellular organisms result. Multicellular bodies have advantages over unicellular ones, e.g. cells can specialize to perform specific biochemical functions, different kinds of structural materials and structures can be produced, and the control center of the cell, the nucleus, is more efficient with fewer activities to control.

Two kinds of cells are: Prokaryotes and eukaryotes. All cells have:

- An outer membrane, the plasma membrane (also called the cell membrane or plasmalemma), composed of a lipid bilayer in which proteins are embedded.
- Genetic material, found in molecules of deoxyribonucleic acid (DNA).

The fundamental separation: Based on cellular structure, the living world can be divided into two kinds of organisms: the prokaryotes and the eukaryotes. The cells of prokaryotes (the bacteria) lack a nucleus and other membrane-bounded cell organelles whereas those of the eukaryotes—the rest of the living world—have nuclei and a complex internal structure with membrane-bounded cell organelles. Table compares features of the cells of the two basic types.

Table: Comparison of prokaryote and eukaryotes.

Features	Prokaryotes	Eukaryotes
Average size of cells	1-10 Φm	10-100 Φm
Nucleus	Absent (have a "nucleoid" region)	Present as a distinct structure.
Nuclear envelope (membrane)	Absent	Present, surround the nucleus.
Genetic material (DNA)	Present in a one large circular molecule, which acts as a "chromosome" in the nucleoid region.	Present in linear chromosome in the nucleus, usually many per cell.
DNA bound to histones (proteins)	No	Yes
Cytoskeleton of microtubules and actin flements	Absent	Present
Membrane – bounded organelles	Present	Present
Ribosomes membrane	Present	Present
Cell wall	Present	Present in plants and some other organisms.

Origin of the eukaryote type: Biologists in general agree that early organisms were prokaryotes that developed ways to survive on a young Earth whose atmosphere and conditions were much different from todays. Descendants of these ancient prokaryotes—the organisms of today—carry in their genes solutions to old problems as well as modifications made over 3.5 billion years in response to changing Earth conditions.

Serial End Symbiotic Theory

The mitochondria and chloroplasts of plants differ from other organelles in their semi-independence. In many ways they resemble bacteria: Both bacteria and the organellescan divide, their DNA is organized similarly, and both have small ribosomes. Additionally, antibiotics inhibit protein synthesis in ribosomes of mitochondria, chloroplasts, and bacteria, but have no effect on synthesis in cytosomal ribosomes of eukaryotic cells. These structural and physiological differences lead some biologists to postulate the serial endosymbiotic theory of organelles in eukaryote cells. (Asymbiosis is the close association and living together of individuals of different species.) The theory suggests that, early in the development of life on Earth, what are now organelles of larger, eukaryotic cells were simple, free-living prokaryotes. At some point in time, each independently developed a symbiotic relationship with larger, heterotrophic cells bringing to the partnership energy-capturing and -converting abilities and receiving in trade a physical home that provided the chemical necessities for life and protection from severe environmental stresses. The combination proved successful. Cells with the combined attributes were able to colonize previously hostile environments and the stage was set for the invasion of land and development of green, multicellular, terrestrial plants. The process occurred repeatedly—it was a serial endosymbiosis.

Cell Cycle

In plants, as in all eukaryotic life-forms, the cell cycle comprises the processes of cell division, constituted by three preparatory phases (G, G, and S), followed by mitosis (nuclear division) and cytokinesis (cytoplasm division).

One of the fundamental characteristics of living organisms is their ability to grow and reproduce. At the cellular level, growth is accomplished by a gain of mass, followed by division into two daughter cells.

In unicellular species, such as bacteria and green algae, this division results in the production of new organisms. In multicellular organisms, such as plants, cells must divide many times to produce new individuals, and additional processes of differentiation into mature cell types must occur.

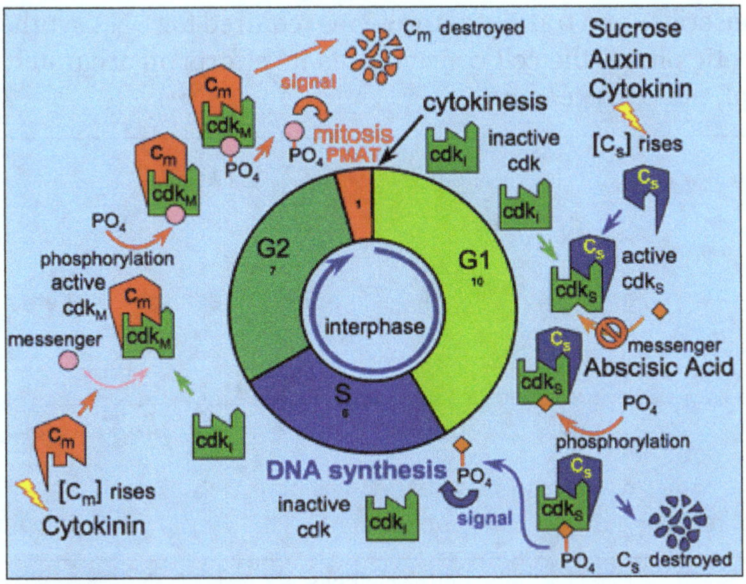

Eukaryotic Cell Division

Plant cells are eukaryotic, meaning that they contain a nucleus and membrane-bound cellular compartments called organelles, such as chloroplasts, the sites of photosynthesis, and the central vacuole used for storage. In order for eukaryotic cell division to occur, cells must accomplish several steps.

First, a cell must reach a sufficient size to ensure that the daughter cells will be large enough to survive. The cell achieves this status by making needed molecules, using synthetic biochemical reactions, and monitoring the outside environment to ensure continued favorable conditions for reproduction.

Next, a cell must replicate all required macromolecules, including the deoxyribonucleic acid (DNA), as well as the organelles. Finally, the cellmust be able to distribute appropriately the cellular contents, including the genetic material, to the daughter cells. The cyclical repetition of these steps comprises the eukaryotic cell division cycle or, more simply, the cell cycle.

Cell Cycle Phases

The general arrangement of the cell cycle was first determined through microscopic observations of living cells undergoing cell division. Originally, four stages were identified: G1 phase, S phase, G2 phase, and M phase. G1 and G2 phases were named as gap phases, because it was presumed that the cell did not appear to accomplish any specific events during these phases.

G1 phase occurs after the previous cell division and is the phase where the cell grows, metabolizes, and prepares to undergo another round of cell division. During G1 phase, the cell also transcribes and translates the genes required for DNA synthesis. In S phase, the DNA synthetic phase, the cell concentrates its efforts on accurately replicating its genetic material, the DNA genome.

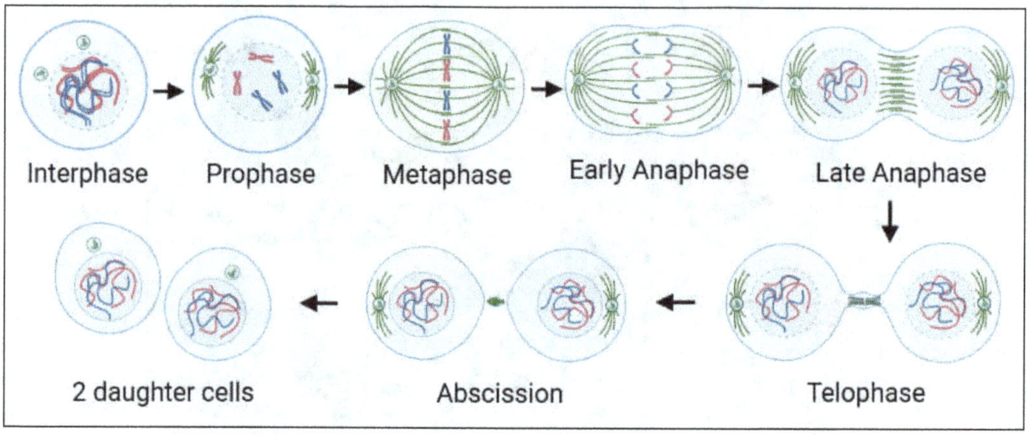

Cell Cycle Phases.

In G2 phase, the cell coordinates both transcription and translation to synthesize proteins that are required specifically for asexual cell division, or mitosis. During M phase, the DNA chromosomes condense and are divided into the two daughter cells. G1, S, and G2 phases together are referred to as interphase.

Mitosis itself is composed of four phases:

- Prophase, in which chromosomes are condensed, homologous chromosomes are paired together, and the spindle apparatus made of microtubules forms.
- Metaphase, in which the paired chromosomes are lined up across the centre of the cell on the metaphase plate.
- Anaphase, in which the homologous chromosomes are pulled to separate poles in the dividing cell by the attached spindle apparatus.
- Telophase, in which the daughter cell chromosomes are collected together at the poles.

Mitosis is followed by cytokinesis, or the process by which the two daughter cells are physically separated. Because plant cells have cell walls, the division of one cell into two daughter cells requires the formation of a cell plate to complete cytokinesis. This cell plate grows outward between the two new nuclei. Once the cell plate reaches the walls of the dividing cell, it forms the cell wall that separates the two new cells.

The cell will die if it goes through any cell cycle phase out of order. Therefore, the cell has evolved sophisticated checkpoints to ensure that critical events, such as DNA replication and cell division, occur in the correct order and that each required step is completed prior to movement to the next phase.

Additionally, external conditions, such as a change in nutrient availability, can cause a cellular checkpoint response. There are two main points in the cell cycle which are regulated by checkpoints, the G1-to-S transition and the G2-to-M transition.

The G1-to-S transition is important because once the cell commits to divide and starts to copy its DNA, it must complete cell division or die. After DNA replication is complete, the G2-to-M transition is important for the cell to make sure that all of the DNA has been copied correctly, or the daughter cells will not have the full complement of genetic material.

CDKs and Cyclins

Because of the nature of the phosphate chemical group, the addition of these groups causes conformational changes in the three-dimensional structure of the proteins that accept them. The cell cycle is controlled by the actions of kinases called cyclin-dependent kinases, or CDKs. Each CDK has a partner called a cyclin, which is required for the kinase activity to place phosphate groups on target proteins.

The CDK/cyclin pairs perform kinase reactions, which activate proteins leading to completion of cell cycle phases. Opposing the action of the CDK and cyclin are cyclin-dependent kinase inhibitors, or CKIs, which act as brakes to prevent CDK/cyclin activity pairs until the cell is ready to go on to the next step in the cell cycle. Many of the cell cycle control mechanisms have been conserved throughout all eukaryotic cells, including the plant cell. Arabidopsis is a genus that has been widely characterized and used as an experimental model for plant growth and differentiation in scientific laboratoriesworldwide.

In plants such as Arabidopsis, two major groups of CDKs have been studied, the A-type and B-type CDKs. The A-type CDKs show kinase activity during the S, G2, and M phases of the cell cycle and regulate the transitions from G1-to-S and G2-to-M phases. The B-type CDKs are active during mitosis and regulate only the G -to-M transition in plants.

Photosynthesis

Plant cells are the building blocks of plants. Photosynthesis is the major function performed by plant cells. Photosynthesis is the process by which green plants and certain other organisms transform light energy into chemical energy. During photosynthesis in green plants, light energy is captured and used to convert water, carbon dioxide, and minerals into oxygen and energy-rich organic compounds.

Requirements for food, materials, and energy in a world where human population is rapidly growing have created a need to increase both the amount of photosynthesis and the efficiency of converting photosynthetic output into products useful to people. One response to those needs—the so-called Green Revolution, begun in the mid-20th century—achieved enormous improvements in agricultural yield through the use of chemical fertilizers, pest and plant-disease control, plant breeding, and mechanized tilling, harvesting, and crop processing. This effort limited severe famines to a few areas of the world despite rapid population growth, but it did not eliminate widespread malnutrition. Moreover, beginning in the early 1990s, the rate at which yields of major crops increased began to decline. This was especially true for rice in Asia. Rising costs associated with sustaining high rates of agricultural production, which required ever-increasing inputs of fertilizers and pesticides and constant development of new plant varieties, also became problematic for farmers in many countries.

Photosynthesis: Diagram of photosynthesis showing how water, light, and carbon dioxide are absorbed by a plant to produce oxygen, sugars, and more carbon dioxide.

A second agricultural revolution, based on plant genetic engineering, was forecast to lead to increases in plant productivity and thereby partially alleviate malnutrition. Since the 1970s, molecular biologists have possessed the means to alter a plant's genetic material (deoxyribonucleic acid, or DNA) with the aim of achieving improvements in disease and drought resistance, product yield and quality, frost hardiness, and other desirable properties. However, such traits are inherently complex, and the process of making changes to crop plants through genetic engineering has turned out to be more complicated than anticipated. In the future such genetic engineering may result in improvements in the process of photosynthesis, but by the first decades of the 21st century, it had yet to demonstrate that it could dramatically increase crop yields.

Another intriguing area in the study of photosynthesis has been the discovery that certain animals are able to convert light energy into chemical energy. The emerald green sea slug (*Elysia chlorotica*), for example, acquires genes and chloroplasts from *Vaucheria litorea*, an alga it consumes, giving it a limited ability to produce chlorophyll. When enough chloroplasts are assimilated, the slug may forgo the ingestion of food. The pea aphid (*Acyrthosiphon pisum*) can harness light to manufacture the energy-rich compound adenosine triphosphate (ATP); this ability has been linked to the aphid's manufacture of carotenoid pigments.

General Characteristics of Photosynthesis

Development of the Idea

The study of photosynthesis began in 1771 with observations made by the English clergyman and scientist Joseph Priestley. Priestley had burned a candle in a closed container until the air within the container could no longer support combustion. He then placed a sprig of mint plant in the container and discovered that after several days the mint had produced some substance (later recognized as oxygen) that enabled the confined air to again support combustion. In 1779 the Dutch physician Jan Ingenhousz expanded upon Priestley's work, showing that the plant had to be exposed to light if the combustible substance (i.e., oxygen) was to be restored. He also demonstrated that this process required the presence of the green tissues of the plant.

In 1782 it was demonstrated that the combustion-supporting gas (oxygen) was formed at the expense of another gas, or "fixed air," which had been identified the year before as carbon dioxide. Gas-exchange experiments in 1804 showed that the gain in weight of a plant grown in a carefully weighed pot resulted from the uptake of carbon, which came entirely from absorbed carbon dioxide, and water taken up by plant roots; the balance is oxygen, released back to the atmosphere. Almost half a century passed before the concept of chemical energy had developed sufficiently to permit the discovery (in 1845) that light energy from the sun is stored as chemical energy in products formed during photosynthesis.

Overall Reaction of Photosynthesis

In chemical terms, photosynthesis is a light-energized oxidation–reduction process. (Oxidation refers to the removal of electrons from a molecule; reduction refers to the gain of electrons by a molecule.) In plant photosynthesis, the energy of light is used to drive the oxidation of water (H_2O), producing oxygen gas (O_2), hydrogen ions (H^+), and electrons. Most of the removed electrons and hydrogen ions ultimately are transferred to carbon dioxide (CO_2), which is reduced to organic products. Other electrons and hydrogen ions are used to reduce nitrate and sulfate to amino and sulfhydryl groups in amino acids, which are the building blocks of proteins. In most green cells, carbohydrates—especially starch and the sugar sucrose—are the major direct organic products of photosynthesis. The overall reaction in which carbohydrates—represented by the general formula (CH_2O)—are formed during plant photosynthesis can be indicated by the following equation:

$$CO_2 + 2H_2O \xrightarrow[\text{green plants}]{\text{light}} (CH_2O) + O_2 + H_2O$$

This equation is merely a summary statement, for the process of photosynthesis actually involves numerous reactions catalyzed by enzymes (organic catalysts). These reactions occur in two stages: the "light" stage, consisting of photochemical (i.e., light-capturing) reactions; and the "dark" stage, comprising chemical reactions controlled by enzymes. During the first stage, the energy of light is absorbed and used to drive a series of electron transfers, resulting in the synthesis of ATP and the electron-donor-reduced nicotine adenine dinucleotide phosphate (NADPH). During the dark stage, the ATP and NADPH formed in the light-capturing reactions are used to reduce carbon dioxide to organic carbon compounds. This assimilation of inorganic carbon into organic compounds is called carbon fixation.

During the 20th century, comparisons between photosynthetic processes in green plants and in certain photosynthetic sulfur bacteria provided important information about the photosynthetic mechanism. Sulfur bacteria use hydrogen sulfide (H_2S) as a source of hydrogen atoms and produce sulfur instead of oxygen during photosynthesis. The overall reaction is:

$$CO_2 + 2H_2S \xrightarrow[\text{sulfur bacteria}]{\text{light}} (CH_2O) + S_2 + H_2O$$

In the 1930s Dutch biologist Cornelis van Niel recognized that the utilization of carbon dioxide to form organic compounds was similar in the two types of photosynthetic organisms. Suggesting that differences existed in the light-dependent stage and in the nature of the compounds used as a source of hydrogen atoms, he proposed that hydrogen was transferred from hydrogen sulfide (in bacteria) or water (in green plants) to an unknown acceptor (called A), which was reduced to H_2A. During the dark reactions, which are similar in both bacteria and green plants, the reduced acceptor (H_2A) reacted with carbon dioxide (CO_2) to form carbohydrate (CH_2O) and to oxidize the unknown acceptor to A. This putative reaction can be represented as:

$$CO_2 + 2H_2A \xrightarrow{\text{light}} (CH_2O) + 2A + H_2O$$

Van Niel's proposal was important because the popular (but incorrect) theory had been that oxygen was removed from carbon dioxide (rather than hydrogen from water, releasing oxygen) and that carbon then combined with water to form carbohydrate (rather than the hydrogen from water combining with CO_2 to form CH_2O).

By 1940 chemists were using heavy isotopes to follow the reactions of photosynthesis. Water marked with an isotope of oxygen (^{18}O) was used in early experiments. Plants that photosynthesized in the presence of water containing $H_2{}^{18}O$ produced oxygen gas containing ^{18}O; those that photosynthesized in the presence of normal water produced normal oxygen gas. These results provided definitive support for van Niel's theory that the oxygen gas produced during photosynthesis is derived from water.

Basic Products of Photosynthesis

As has been stated, carbohydrates are the most-important direct organic product of photosynthesis in the majority of green plants. The formation of a simple carbohydrate, glucose, is indicated by a chemical equation:

$$\underset{\text{carbon dixode}}{6CO_2} + \underset{\text{water}}{12H_2O} \xrightarrow[\text{green plants}]{\text{light}} \underset{\text{glucose}}{(C_6H_{12}O_6)} + \underset{\text{oxygen}}{6O_2} + \underset{\text{water}}{6H_2O}$$

Little free glucose is produced in plants; instead, glucose units are linked to form starch or are joined with fructose, another sugar, to form sucrose.

Not only carbohydrates, as was once thought, but also amino acids, proteins, lipids (or fats), pigments, and other organic components of green tissues are synthesized during photosynthesis. Minerals supply the elements (e.g., nitrogen, N; phosphorus, P; sulfur, S) required to form these compounds. Chemical bonds are broken between oxygen (O) and carbon (C), hydrogen (H), nitrogen, and sulfur, and new bonds are formed in products that include gaseous oxygen (O_2) and organic compounds. More energy is required to break the bonds between oxygen and other elements (e.g., in water, nitrate, and sulfate) than is released when new bonds form in the products. This difference in bond energy accounts for a large part of the light energy stored as chemical energy in the organic products formed during photosynthesis. Additional energy is stored in making complex molecules from simple ones.

Evolution of the Process

Although life and the quality of the atmosphere today depend on photosynthesis, it is likely that green plants evolved long after the first living cells. When Earth was young, electrical storms and solar radiation probably provided the energy for the synthesis of complex molecules from abundant simpler ones, such as water, ammonia, and methane. The first living cells probably evolved from these complex molecules (see life: Production of polymers). For example, the accidental joining (condensation) of the amino acid glycine and the fatty acid acetate may have formed complex organic molecules known as porphyrins. These molecules, in turn, may have evolved further into coloured molecules called pigments—e.g., chlorophylls of green plants, bacteriochlorophyll of photosynthetic bacteria, hemin (the red pigment of blood), and cytochromes, a group of pigment molecules essential in both photosynthesis and cellular respiration.

Primitive coloured cells then had to evolve mechanisms for using the light energy absorbed by their pigments. At first, the energy may have been used immediately to initiate reactions useful to the cell. As the process for utilization of light energy continued to evolve, however, a larger part of the absorbed light energy probably was stored as chemical energy, to be used to maintain life. Green plants, with their ability to use light energy to convert carbon dioxide and water to carbohydrates and oxygen, are the culmination of this evolutionary process.

The first oxygenic (oxygen-producing) cells probably were the blue-green algae (cyanobacteria), which appeared about two billion to three billion years ago. These microscopic organisms are believed to have greatly increased the oxygen content of the atmosphere, making possible the development of aerobic (oxygen-using) organisms. Cyanophytes are prokaryotic cells; that is, they contain no distinct membrane-enclosed subcellular particles (organelles), such as nuclei and chloroplasts. Green plants, by contrast, are composed of eukaryotic cells, in which the photosynthetic apparatus is contained within membrane-bound chloroplasts. The complete genome sequences of cyanobacteria and higher plants provide evidence that the first photosynthetic eukaryotes were likely the red algae that developed when nonphotosynthetic eukaryotic cells engulfed cyanobacteria. Within the host cells, these cyanobacteria evolved into chloroplasts.

There are a number of photosynthetic bacteria that are not oxygenic (e.g., the sulfur bacteria. The evolutionary pathway that led to these bacteria diverged from the one that resulted in oxygenic organisms. In addition to the absence of oxygen production, nonoxygenic photosynthesis differs from oxygenic photosynthesis in two other ways: light of longer wavelengths is absorbed and used by pigments called bacteriochlorophylls, and reduced compounds other than water (such as hydrogen sulfide or organic molecules) provide the electrons needed for the reduction of carbon dioxide.

Factors that Influence the Rate of Photosynthesis

The rate of photosynthesis is defined in terms of the rate of oxygen production either per unit mass (or area) of green plant tissues or per unit weight of total chlorophyll. The amount of light, the carbon dioxide supply, temperature, water supply, and the availability of minerals are the most important environmental factors that affect the rate of photosynthesis in land plants. The rate of photosynthesis is also determined by the plant species and its physiological state—e.g., its health, its maturity, and whether it is in flower.

Light Intensity and Temperature

The complex mechanism of photosynthesis includes a photochemical, or light-harvesting, stage and an enzymatic, or carbon-assimilating, stage that involves chemical reactions. These stages can be distinguished by studying the rates of photosynthesis at various degrees of light saturation (i.e., intensity) and at different temperatures. Over a range of moderate temperatures and at low to medium light intensities (relative to the normal range of the plant species), the rate of photosynthesis increases as the intensity increases and is relatively independent of temperature. As the light intensity increases to higher levels, however, the rate becomes saturated; light "saturation" is achieved at a specific light intensity, dependent on species and growing conditions. In the light-dependent range before saturation, therefore, the rate of photosynthesis is determined by the rates of photochemical steps. At high light intensities, some of the chemical reactions of the dark stage become rate-limiting. In many land plants, a process called photorespiration occurs, and its influence upon photosynthesis increases with rising temperatures. More specifically, photorespiration competes with photosynthesis and limits further increases in the rate of photosynthesis, especially if the supply of water is limited.

Carbon Dioxide

Included among the rate-limiting steps of the dark stage of photosynthesis are the chemical reactions by which organic compounds are formed by using carbon dioxide as a carbon source. The rates of these reactions can be increased somewhat by increasing the carbon dioxide concentration. Since the middle of the 19th century, the level of carbon dioxide in the atmosphere has been rising because of the extensive combustion of fossil fuels, cement production, and land-use changes associated with deforestation. The atmospheric level of carbon dioxide climbed from about 0.028 percent in 1860 to 0.032 percent by 1958 (when improved measurements began) and to 0.041 percent by 2020. This increase in carbon dioxide directly increases plant photosynthesis up to a point, but the size of the increase depends on the species and physiological condition of the plant. Furthermore, most scientists maintain that increasing levels of atmospheric carbon dioxide affect climate, increasing global temperatures and changing rainfall patterns. Such changes will also affect photosynthesis rates.

Water

For land plants, water availability can function as a limiting factor in photosynthesis and plant growth. Besides the requirement for a small amount of water in the photosynthetic reaction itself, large amounts of water are transpired from the leaves; that is, water evaporates from the leaves to the atmosphere via the stomata. Stomata are small openings through the leaf epidermis, or outer skin; they permit the entry of carbon dioxide but inevitably also allow the exit of water vapour. The stomata open and close according to the physiological needs of the leaf. In hot and arid climates the stomata may close to conserve water, but this closure limits the entry of carbon dioxide and hence the rate of photosynthesis. The decreased transpiration means there is less cooling of the leaves and hence leaf temperatures rise. The decreased carbon dioxide concentration inside the leaves and the increased leaf temperatures favour the wasteful process of photorespiration. If the level of carbon dioxide in the atmosphere increases, more carbon dioxide could enter through a smaller opening of the stomata, so more photosynthesis could occur with a given supply of water.

Minerals

Several minerals are required for healthy plant growth and for maximum rates of photosynthesis. Nitrogen, sulfate, phosphate, iron, magnesium, calcium, and potassium are required in substantial amounts for the synthesis of amino acids, proteins, coenzymes, deoxyribonucleic acid (DNA) and ribonucleic acid (RNA), chlorophyll and other pigments, and other essential plant constituents. Smaller amounts of such elements as manganese, copper, and chloride are required in photosynthesis. Some other trace elements are needed for various nonphotosynthetic functions in plants.

Internal Factors

Each plant species is adapted to a range of environmental factors. Within this normal range of conditions, complex regulatory mechanisms in the plant's cells adjust the activities of enzymes (i.e., organic catalysts). These adjustments maintain a balance in the overall photosynthetic process and control it in accordance with the needs of the whole plant. With a given plant species, for example, doubling the carbon dioxide level might cause a temporary increase of nearly twofold in the rate of photosynthesis; a few hours or days later, however, the rate might fall to the original level because photosynthesis produced more sucrose than the rest of the plant could use. By contrast, another plant species provided with such carbon dioxide enrichment might be able to use more sucrose, because it had more carbon-demanding organs, and would continue to photosynthesize and to grow faster throughout most of its life cycle.

Energy Efficiency of Photosynthesis

The energy efficiency of photosynthesis is the ratio of the energy stored to the energy of light absorbed. The chemical energy stored is the difference between that contained in gaseous oxygen and organic compound products and the energy of water, carbon dioxide, and other reactants. The amount of energy stored can only be estimated because many products are formed, and these vary with the plant species and environmental conditions. If the equation for glucose formation given earlier is used to approximate the actual storage process, the production of one mole (i.e., 6.02×10^{23} molecules; abbreviated *N*) of oxygen and one-sixth mole of glucose results in the storage of about 117 kilocalories (kcal) of chemical energy. This amount must then be compared with the energy of light absorbed to produce one mole of oxygen in order to calculate the efficiency of photosynthesis.

Light can be described as a wave of particles known as photons; these are units of energy, or light quanta. The quantity *N* photons is called an einstein. The energy of light varies inversely with the length of the photon waves; that is, the shorter the wavelength, the greater the energy content. The energy (e) of a photon is given by the equation $e = hc/\lambda$, where c is the velocity of light, h is Planck's constant, and λ is the light wavelength. The energy (E) of an einstein is $E = Ne = Nhc/\lambda = 28{,}600/\lambda$, when E is in kilocalories and λ is given in nanometres (nm; 1 nm = 10^{-9} metres). An einstein of red light with a wavelength of 680 nm has an energy of about 42 kcal. Blue light has a shorter wavelength and therefore more energy than red light. Regardless of whether the light is blue or red, however, the same number of einsteins are required for photosynthesis per mole of oxygen formed. The part of the solar spectrum used by plants has an estimated mean wavelength of 570 nm; therefore, the energy of light used during photosynthesis is approximately 28,600/570, or 50 kcal per einstein.

In order to compute the amount of light energy involved in photosynthesis, one other value is needed: the number of einsteins absorbed per mole of oxygen evolved. This is called the quantum requirement. The minimum quantum requirement for photosynthesis under optimal conditions is about nine. Thus, the energy used is 9 × 50, or 450 kcal per mole of oxygen evolved. Therefore, the estimated maximum energy efficiency of photosynthesis is the energy stored per mole of oxygen evolved, 117 kcal, divided by 450—that is, 117/450, or 26 percent.

The actual percentage of solar energy stored by plants is much less than the maximum energy efficiency of photosynthesis. An agricultural crop in which the biomass (total dry weight) stores as much as 1 percent of total solar energy received on an annual areawide basis is exceptional, although a few cases of higher yields (perhaps as much as 3.5 percent in sugarcane) have been reported. There are several reasons for this difference between the predicted maximum efficiency of photosynthesis and the actual energy stored in biomass. First, more than half of the incident sunlight is composed of wavelengths too long to be absorbed, and some of the remainder is reflected or lost to the leaves. Consequently, plants can at best absorb only about 34 percent of the incident sunlight. Second, plants must carry out a variety of physiological processes in such nonphotosynthetic tissues as roots and stems; these processes, as well as cellular respiration in all parts of the plant, use up stored energy. Third, rates of photosynthesis in bright sunlight sometimes exceed the needs of the plants, resulting in the formation of excess sugars and starch. When this happens, the regulatory mechanisms of the plant slow down the process of photosynthesis, allowing more absorbed sunlight to go unused. Fourth, in many plants, energy is wasted by the process of photorespiration. Finally, the growing season may last only a few months of the year; sunlight received during other seasons is not used. Furthermore, it should be noted that if only agricultural products (e.g., seeds, fruits, and tubers, rather than total biomass) are considered as the end product of the energy-conversion process of photosynthesis, the efficiency falls even further.

Chloroplasts, The Photosynthetic Units of Green Plants

The process of plant photosynthesis takes place entirely within the chloroplasts. Detailed studies of the role of these organelles date from the work of British biochemist Robert Hill. About 1940 Hill discovered that green particles obtained from broken cells could produce oxygen from water in the presence of light and a chemical compound, such as ferric oxalate, able to serve as an electron acceptor. This process is known as the Hill reaction. During the 1950s Daniel Arnon and other American biochemists prepared plant cell fragments in which not only the Hill reaction but also the synthesis of the energy-storage compound ATP occurred. In addition, the coenzyme NADP was used as the final acceptor of electrons, replacing the nonphysiological electron acceptors used by Hill. His procedures were refined further so that small individual pieces of isolated chloroplast membranes, or lamellae, could perform the Hill reaction. These small pieces of lamellae were then fragmented into pieces so small that they performed only the light reactions of the photosynthetic process. It is now possible also to isolate the entire chloroplast so that it can carry out the complete process of photosynthesis, from light absorption, oxygen formation, and the reduction of carbon dioxide to the formation of glucose and other products.

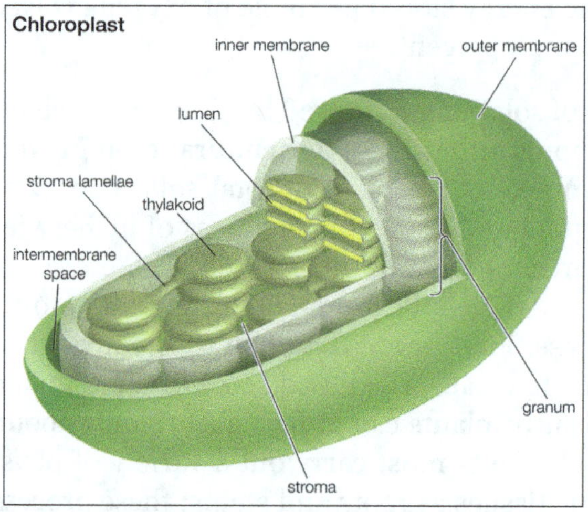

Chloroplast structure: The internal (thylakoid) membrane vesicles are organized into stacks, which reside in a matrix known as the stroma. All the chlorophyll in the chloroplast is contained in the membranes of the thylakoid vesicles.

Structural Features

The intricate structural organization of the photosynthetic apparatus is essential for the efficient performance of the complex process of photosynthesis. The chloroplast is enclosed in a double outer membrane, and its size approximates a spheroid about 2,500 nm thick and 5,000 nm long. Some single-celled algae have one chloroplast that occupies more than half the cell volume. Leaf cells of higher plants contain many chloroplasts, each approximately the size of the one in some algal cells.

When thin sections of a chloroplast are examined under the electron microscope, several features are apparent. Chief among these are the intricate internal membranes (i.e., the lamellae) and the stroma, a colourless matrix in which the lamellae are embedded. Also visible are starch granules, which appear as dense bodies.

The stroma is basically a solution of enzymes and small molecules. The dark reactions occur in the stroma, the soluble enzymes of which catalyze the conversion of carbon dioxide and minerals to carbohydrates and other organic compounds. The capacity for carbon fixation and reduction is lost if the outer membrane of the chloroplast is broken, allowing the stroma enzymes to leak out.

A single lamella, which contains all the photosynthetic pigments, is approximately 10–15 nm thick. The lamellae exist in more-or-less flat sheets, a few of which extend through much of the length of the chloroplast. Examination of cross sections of lamellae under the electron microscope shows that their edges are joined to form closed hollow disks that are called thylakoids ("saclike"). The chloroplasts of most higher plants have regions, called grana, in which the thylakoids are very tightly stacked. When viewed by electron microscopy at an oblique angle, the grana appear as stacks of disks. When viewed in cross section, it is apparent that some thylakoids extend from one grana through the stroma into other grana. The thin aqueous spaces inside the thylakoids are believed to be connected with each other via these stroma thylakoids. These thylakoid spaces are isolated from the stroma spaces by the relatively impermeable lamellae.

The light reactions occur exclusively in the thylakoids. The complex structural organization of lamellae is required for proper thylakoid function; intact thylakoids are necessary for the formation of ATP. Thylakoids that have been broken down to smaller units can no longer form ATP, even when the conversion of light into chemical energy occurs during electron transport in these units. Such lamellar fragments can carry out the Hill reaction, with the transfer of electrons from water to $NADP^+$.

Chemical Composition of Lamellae

Lipids

Lamellae consist of about equal amounts of lipids and proteins. About one-fourth of the lipid portion of the lamellae consists of pigments and coenzymes; the remainder consists of various lipids, including polar compounds such as phospholipids and galactolipids. These polar lipid molecules have "head" groups that attract water (i.e., are hydrophilic) and fatty acid "tails" that are oil soluble and repel water (i.e., are hydrophobic). When polar lipids are placed in an aqueous environment, they can line up with the fatty acid tails side by side. A second layer of phospholipids forms tail-to-tail with the first, establishing a lipid bilayer in which the hydrophilic heads are in contact with the aqueous solution on each side of the bilayer. Sandwiched between the heads are the hydrophobic tails, creating a hydrophobic environment from which water is excluded. This lipid bilayer is an essential feature of all biological membranes. The hydrophobic parts of proteins and lipid-soluble cofactors and pigments are dissolved or embedded in the lipid bilayer. Lamellar membranes can function as electrical insulating material and permit a charge, or potential difference, to develop across the membrane. Such a charge can be a source of chemical or electrical energy.

Approximately one-fifth of the lamellar lipids are chlorophyll molecules; one type, chlorophyll *a*, is more abundant than the second type, chlorophyll *b*. The chlorophyll molecules are specifically bound to small protein molecules. Most of these chlorophyll-proteins are "light-harvesting" pigments. These absorb light and pass its energy on to special chlorophyll *a* molecules that are directly involved in the conversion of light energy to chemical energy. When one of these special chlorophyll *a* molecules is excited by light energy (as described later), it gives up an electron. There are two types of these special chlorophyll *a* molecules: one, called P_{680}, has an absorption spectrum that peaks at 684 nm; the other, called P_{700}, shows an absorption peak at 700 nm.

Although chlorophylls are the main light-absorbing molecules in green plants, there are other pigments such as carotenes and carotenoids (which are responsible for the yellow-orange colour of carrots). Carotenes can also absorb light and may supplement chlorophyll as the light-absorbing molecules in some plant cells. The light energy absorbed by carotenes must be passed to chlorophyll before conversion to chemical energy can occur. Carotenoids are part of a cycle that renders excess energy beyond the level of light saturation harmless, effectively serving as "lightning rods" in the process.

Proteins

Many of the lamellar proteins are components of the chlorophyll–protein complexes. Other proteins include enzymes and protein-containing coenzymes. Enzymes are required as organic catalysts for specific reactions within the lamellae. Protein coenzymes, also called cofactors, include important electron carrier molecules called cytochromes, which are iron-containing pigments with the pigment portions attached to protein molecules. During electron transfer, an electron is accepted by an iron atom in the pigment portion of a cytochrome molecule, which thus is reduced; then the electron is transferred to the iron atom in the next cytochrome carrier in the electron transfer chain, thus oxidizing the first cytochrome and reducing the next one in the chain.

In addition to the metal atoms found in the pigment portions of cytochrome molecules, metal atoms also are found in other protein molecules of the lamellae. In proteins with a total molecular weight of 900,000 (based on the weight of hydrogen as one), there are 2 atoms of manganese, 10 atoms of iron, and 6 atoms of copper. These metal atoms are required for the catalytic activity of some of the enzymes important in photosynthesis. The manganese atoms are involved in water-splitting and oxygen formation. Both copper- and iron-containing proteins function in electron transport between water and the final electron-acceptor molecule of the light stage of photosynthesis, an iron-containing protein called ferredoxin. Ferredoxin is a soluble component in the chloroplasts. In its reduced form, it gives electrons directly to the systems that reduce nitrate and sulfate and via NADPH to the system that reduces carbon dioxide. A copper-containing protein called plastocyanin (PC) carries electrons at one point in the electron transport chain. PC molecules are water soluble and can move through the inner space of the thylakoids, carrying electrons from one place to another.

Quinones

Small molecules called plastoquinones are found in substantial numbers in the lamellae. Like the cytochromes, quinones have important roles in carrying electrons between the components of the light reactions. Since they are lipid soluble, they can diffuse through the membrane. They can carry one or two electrons, and, in their reduced form (with added electrons), they carry hydrogen atoms that can be released as hydrogen ions when the added electrons are passed on, for example, to a cytochrome.

The Process of Photosynthesis: The Light Reactions

Light Absorption and Energy Transfer

The light energy absorbed by a chlorophyll molecule excites some electrons within the structure of the molecule to higher energy levels, or excited states. Light of shorter wavelength (such as blue) has more energy than light of longer wavelength (such as red), so absorption of blue light creates an excited state of higher energy. A molecule raised to this higher energy state quickly gives up the "extra" energy as heat and falls to its lowest excited state. This lowest excited state is similar to that of a molecule that has just absorbed the longest wavelength light capable of exciting it. In the case of chlorophyll *a*, this lowest excited state corresponds to that of a molecule that has absorbed red light of about 680 nm.

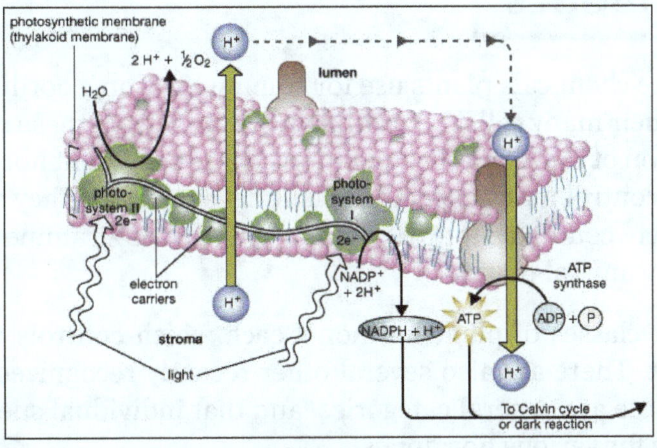

Light reaction.

The light reaction of photosynthesis. The light reaction occurs in two photosystems (units of chlorophyll molecules). Light energy (indicated by wavy arrows) absorbed by photosystem II causes the formation of high-energy electrons, which are transferred along a series of acceptor molecules in an electron transport chain to photosystem I. Photosystem II obtains replacement electrons from water molecules, resulting in their split into hydrogen ions (H^+) and oxygen atoms. The oxygen atoms combine to form molecular oxygen (O_2), which is released into the atmosphere. The hydrogen ions are released into the lumen. Additional hydrogen ions are pumped into the lumen by electron acceptor molecules. This creates a high concentration of ions inside the lumen. The flow of hydrogen ions back across the photosynthetic membrane provides the energy needed to drive the synthesis of the energy-rich molecule adenosine triphosphate (ATP). High-energy electrons, which are released as photosystem I absorbs light energy, are used to drive the synthesis of nicotine adenine dinucleotide phosphate

(NADPH). Photosystem I obtains replacement electrons from the electron transport chain. ATP provides the energy and NADPH provides the hydrogen atoms needed to drive the subsequent photosynthetic dark reaction, or Calvin cycle.

The return of a chlorophyll *a* molecule from its lowest excited state to its original low-energy state (ground state) requires the release of the extra energy of the excited state. This can occur in one of several ways. In photosynthesis, most of this energy is conserved as chemical energy by the transfer of an electron from a special chlorophyll *a* molecule (P_{680} or P_{700}) to an electron acceptor. When this electron transfer is blocked by inhibitors, such as the herbicide dichlorophenylmethylurea (DCMU), or by low temperature, the energy can be released as red light. Such reemission of light is called fluorescence. The examination of fluorescence from photosynthetic material in which electron transfer has been blocked has proved to be a valuable tool for scientists studying the light reactions.

Plant Hormones

Plant hormones are chemicals plants use for communication, coordination, and development between their many cells. Like animals, plants rely on these chemical signals to direct the expression of DNA and the operations of the cell. Plant hormones are natural substances which control many aspects of plant development. They control everything from the length between nodes on the branches to the programmed death, or senescence seen in many annual plants.

There are 5 major classes of plant hormone, each which controls various aspects of plant development. There are also several other recently recognized plant hormones. Remember that these are general categories, and that individual species may have developed novel uses for various hormones.

Types of Plant Hormones

Abscisic Acid

The original name of abscisic acid was dormin, because the plant hormones are heavily involved in the dormancy process. Today, these plant hormones have two main recognized functions in plants. First, they regulate the process of seed development. This helps transform the embryo into a fully-fledged seedling. Second, these plant hormones play a crucial role in the plant's response to temperature and water loss.

As the temperature increases, more water evaporates out of the stoma, little holes in the leaves. As the temperature reaches a point which starts dramatic water loss, abscisic acid is produced and released into the leaves. This causes the stoma to close, and the water is retained within the leaves. Without these plant hormones, plants could not regulate their water content. This is an important and necessary function of vascular plants.

Auxins

Auxins are a class of plant hormones responsible for various aspects of plant growth. Typically, they affect cell enlargement and elongation. They also allow the plant to react to sunlight and gravity, known as phototropism and geotropism, respectively. In many plants auxins are responsible for establishing the apical meristem, and establish the growth direction of the plant.

For this purpose, auxins are typically distributed in a pattern where they are concentrated in the shoots, and less concentrated towards the roots. The concentration of these plant hormones also directs the development in different parts of the plant, depending on the species. In fruit trees, auxins are involved in initiating flowing and finishing the fruiting process. In potatoes and carrots, auxins are involved in the regulation and storage of starches in the roots.

There are several synthetic auxins molecules, created in a laboratory, which can serve as plant hormones. These are often referred to as plant growth regulators. Synthetic auxins have many commercial uses. They can initiate rooting, by activating new growth. They are also used as a weed killer. Synthetic auxins can disrupt the growth cycle of many plants, killing them off. Further, synthetic auxins can be used to culture new plants from tissues, or stop the growth of unwanted branches on ornamental trees. As it was one of the first of the plant hormones discovered, the uses of auxins have expanded greatly.

Cytokinins

The cytokinins are a group of plant hormones which interact directly with the auxins. In doing so, they direct cell differentiation and various aspects of cell metabolism. The cytokinins interact with the plant's DNA, making it express or hide various proteins. This directs cell differentiation within the plant, allowing the plant to develop different tissues for different purposes.

Unlike the auxins, the cytokinins are most concentrated in the roots, and become less concentrated towards the shoots. This counterbalance to auxin allows the plant to develop and maintain an axis, and grow in both directions. Cytokinins applied without auxins produce roots, auxins alone produce buds, and combined they tend to from undifferentiated growth. While there have been many experiments done with these plant hormones, there are not a lot of applications commercially.

Ethylene

Unlike the other plant hormones, ethylene is a single chemical. It is in the form of a gas at regular temperatures, and allows plants a fast way to communicate between cells and other plants. In the early 1900s, it was found that ethylene gas would start the ripening process in fruits. Ethylene is typically found in plants any time damage is incurred. When a stem is bent, bruised, or broken, ethylene is released. As a gas, ethylene quickly diffuses through the fluids of the plant, and can travel through the air. Plants use this hormone to communicate damage to other plants, stimulating them to ripen their fruit or to develop defenses against herbivores.

Ethylene was discovered in the 1960s, and many commercial applications have been developed. As a gas, ethylene can be blown over a crop, stimulating the entire crop to ripen at the same time. This allows commercial farmers to harvest an entire crop at the same time. Ethylene induces fruits, nuts, and vegetables to finish growing and to detach from the stem. This ensures easy collection. These plant hormones are also used to change the sex expression of certain plants, allowing growers to manipulate their crop.

Gibberellins

Gibberellins, like the auxins, are regulatory plant hormones. They, to a greater extent than the auxins, control cell division and overall plant growth. Dwarf plants often have a genetic defect in which gibberellins cannot be produced or utilized. A dwarf plant exposed to extra gibberellins will grow to a normal size. Gibberellins are also responsible for activating a number of enzymes.

Some plants use gibberellins as sexual hormones, helping drive the development of male and female flowers. Along with the auxins, gibberellin plant hormones affect senescence of plant parts. Gibberellins also have an important role in bringing seeds out of dormancy. Gibberellins in the seed activate enzymes such as amylases, which will break down starches to glucose and provide the embryo with energy. The plant hormones also activate other enzymes, which provide the embryo with amino acids and lipids to grow.

In commercial farming, these plant hormones have many uses. Gibberellins are used to increase the size of grapes and other fruit, if applied at the right stage. As gibberellins naturally stimulate seed germination, synthetic ones can also promote seeds to germinate. This can help ensure that all seeds sprout and become viable. Commercial farmers can also use the application of gibberellins to help promote male or female flowers, giving them the ability to selectively breed many plants. Commercial applications of gibberellins are often collected from bacteria which are grown to create the gibberellins.

Other Plant Hormones

There are four more basic plant hormone groups which have been discovered in recent years. The brassinolides are steroid hormones, similar to estrogen and testosterone.

These plant hormones have some function in cell division, while it is not entirely clear how they act. Salicylic acid, another recently discovered hormone, acts like ethylene and allows plants to communicate between individuals. These plant hormones react to pathogens and attack, like an immune system hormone. Another class, the jasmonates, represent a similar plant hormone. Lastly, systemin is a plant defense hormone class involved in activating the defense genes of various plants after a portion of their system has been injured.

References

- Plant-cell, science: britannica.com, Retrieved 11, July 2020
- Plantcell, cells: micro.magnet.fsu.edu, Retrieved 17, January 2020
- Plant-cells-and-tissues, science-education-11091: jove.com, Retrieved 25, August 2020
- Cell-theory, cells, plant-biology, biology: cliffsnotes.com, Retrieved 07, March 2020
- Cell-cycle: lifeofplant.blogspot.com, Retrieved 19, June 2020
- Photosynthesis, science: britannica.com, Retrieved 14, February 2020
- Plant-hormones: biologydictionary.net, Retrieved 21, May 2020

Chapter 2

Types of Plant Cells

Plants cells are eukaryotic cells that have cell walls composed of cellulose, hemicellulose and pectin. Specific plant cells perform specific function in plants. For example, meristemic cells are totipotent and perform continued cell division, and collenchyma cells provide support and structure. This chapter covers different types of cells such as meristematic cells, parenchyma cells, collenchyma cells, sclerenchyma cells, epidermal cells, xylem cells and phloem cells.

Meristematic Cells

Meristem cells are a group of cells that reside at the shoot and root tips of plants. As undifferentiated (or slightly differentiated cells) they are considered as stem cells given that they are the origin of many of the cells that go on to rapidly differentiate/specialize and form various parts of the plant.

While meristem cells share a number of characteristics with stem cells of animals, one of the main differences between the two lies in the fact that meristem cells can be restored and continue dividing which in turn allows indeterminate growth in plants (as long as the necessary resourses are available).

Collectively, meristem cells make up three types of meristematic tissues that include:

- Apical meristems.
- Lateral meristems.
- Basal/Intercalary meristems (present in some plants).

Types, Characteristics and Function

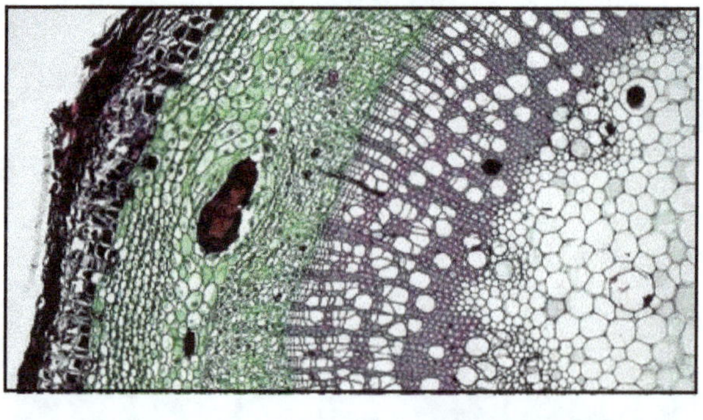

Meristem cells are classified based on their origin and location in the plant. For this reason, there are two major classifications. Different types of these cells have various characteristics and functions that contribute to the growth and development of the plant.

Classification based on Origin

Promeristem Cells

Also known as the primordial meristem or the procambrian meristem, these are the earliest and youngest meristem cells that originate from the embryo. As such, promeristem cells form the first meristematic tissues in the root tips and shoot tip of the plant.

In addition to being very few in number (given that they originate from the embryo and are found in very young plants), these cells are undifferentiated. However, they ultimately go through cell division to produce primary meristem cells.

Primary Meristem Cells

Primary meristematic cells arise from the promeristem and make up the apical tip (present at the shoot tip) as well as the root promeristem. They are therefore more in numbers, as compared to the promeristematic cells and play an important role as the origin of primary tissues (primary growth):

- With actively dividing cells, the primary meristem tissue produces the cortex, epidermis, pith as well as the leaves of a plant.

Secondary Meristem Cells

These cells are produced by primary meristematic cells that are located in the shoot apices and plant roots. Unlike the previous types of cells, secondary meristematic cells are produced once the plant has already started developing. They are involved in the production of secondary tissues including vascular bundles (xylem and phloem):

- Another group of cells known as tertiary meristematic cells in the tertiary meristem have been shown to develop from parenchymatic tissues. These cells reside in the cortex and vascular tissue of plants.
- In plants, primary and secondary meristematic cells contribute to primary and secondary growth of the plant. Apical meristematic cells in the roots and shoot contribute to the primary growth that results in the plant growing longer while the vascular cambium and cork cambium that make up the lateral meristems contribute to the secondary growth (making the plant wider).

Classification based on their Position/Location

Apical Meristematic Cells

Cells of the apical meristem are located at the growing points of the plant. As such, they are present at the shoot, roots as well as branches of the plant. In these locations, they contribute to the length of the plant.

During division, cells of the apical meristem produce new meristematic cells that reside in the shoot tip and roots. Some of the new cells, however, differentiate to produce specialized cells that form different tissues of the plant.

Using cell to cell interactions as well as hormones that act as positional cues, cells of the apical meristem are not only capable of specializing to specific functions (thus forming specific tissues) but also settling in certain parts of the plant. Through the positional cues, certain genes are activated or inhibited thus regulating the differentiation pattern.

The apical meristem is divided in to (SAM) shoot apical meristem (cells located at the tip of branches and plant tip) and the (RAM) root apical meristem where cells are located at the tip of each root. The primary meristem is basal to the shoot apical meristem (SAM) and is composed of cells that are considered to be in their embryonic stage.

These meristematic cells are divided into the following parts:

Protoderm

The differentiation of the protoderm is one of the major events of embryo development. Here, cells of the protodermal cell layer start differentiating following the polarity of the apical meristem and before the meristems form at the opposite ends of the embryo.

This differentiation results in the production of epidermal cells and consequently the epidermis. In the roots, some of the cells elongate (elongation of the cell walls) and form the root hairs. Although the cells formed here have a thin cell wall, they contain cellulose and pectic substances that help protect cells of the roots.

In addition, protoderm cells continually differentiate to produce new epidermal cells of the roots that ensure that new root hairs continue to be formed given that they have a short lifespan (a few days).

In the roots, protoderm plays an important role in the formation of root hairs that are involved in the absorption of nutrients and water in their environment. The epidermis (which is, for the most part, a single cell layer) also covers all organs in the stem of plants thereby acting like a protective layer.

Ground Meristem

Although cells of the ground meristem are a type of primary meristem, as is the case with protoderm cells and cells of the primary procambial, they are segregated and thus set apart from the other cells. Division of these cells results in the production of the cortex, pith as well as a number of other related tissues.

Cells of the ground meristem, therefore, contribute to the growth and development of the plant through the formation of such parts as the root cortex. The endodermis, located in the inner layer of the cortex helps regulate the accumulation of minerals in the roots and thus to other parts of the plant. As the origin of ground tissue systems, cells of the ground meristem are involved in the production of three main types of cells that include:

- Parenchyma: Although they may take up a variety of shapes, most parenchyma cells are spherical or elongated in shape. They may also have two cell walls (a thin primary wall as well as a secondary wall) consisting of a polylayer known as lignin.
    - These cells play a number of important roles related to metabolism (e.g. respiration, secretion, and storage among others). For instance, while the pith is the site of storage for most plants, the mesophyll in leaves is the site of photosynthesis.
    - Parenchyma cells are also capable of being reprogrammed to form different types of cells when need be.
- Collenchyma: Cells of the collenchyma cells serve to provide support to young stems and leaves when they differentiate. In young stems, they can be found beneath the epidermis (as the outer cells of cortex).
    - With regards to structure, collenchyma cells are elongated whose walls consist of cellulose and pectin. Thickening observed at the corner of their cell walls contributes to their supportive function.

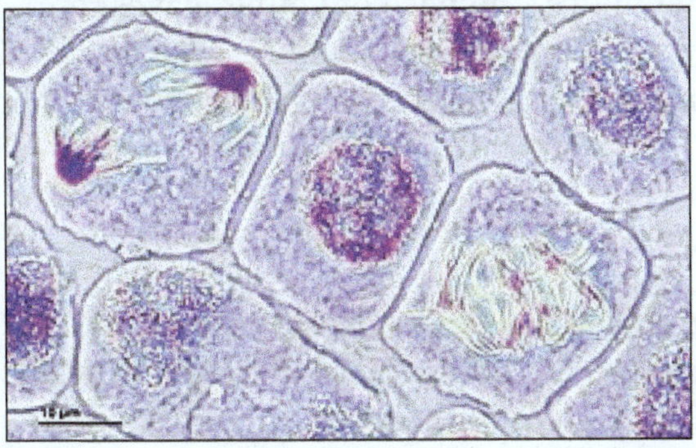

- Sclerenchyma: Like collenchyma cells, cells that make up the sclerenchyma also serve to provide supportive functions. As such, they also have a thicker secondary cell wall that consists of lignin.
    - They include fibers that are elongated cells which act like strengthening

cables and sclereids of varying shapes that either exist as solitary cells or in clusters. Compared to the fibers, sclereids have thicker cell walls that help support the weight of plant organs.

- Cells of the Sclerenchyma die when they mature which causes them to lose their flexibility. As a result, they cannot provide the flexible support observed in collenchyma cells.

Procambium

The procambrian is composed of elongated cells that are characterized by a large nucleus, dense cytoplasm, and proplastids. Differentiation of these cells results in the production of the vascular tissue of the plant (primary xylem and phloem in plants). Various types of vessels develop at different stages and serve different functions. These include:

- Protoxylem: Involved in the transportation of water as the plant roots elongate. The presence of annular rings in the protoxylem also provides supportive function.
- Metaxylem: Found in elongated parts of the roots and are involved in the exchange of water and minerals.
- Phloem cells: Located between the arms of the protoxylem and form the phloem. They are involved in the transportation of food material during the adult life of the plant.

Intercalary Meristematic Cells

Cells of the intercalary meristem are located between the segments of non-meristematic organs (e.g. above nodes, the base of young leaves, etc). These cells are said to originate from primary meristematic cells and tend to lose their meristematic nature as they gradually become permanent. These cells are found in monocot plants such as grass (bamboo etc).

Lateral Meristematic Cells

These cells are located parallel to the sides of organs (lateral side/along the stem) where they contribute to the girth/thickness of the plant. Cells that make up the lateral meristem are characterized by their rectangular shape that develops along one plane:

- Secondary tissue produced by cells of the lateral meristem includes the vascular cambium and cork cambium in the epidermis.
- Cells of the cork cambium replace those of the epidermis in the stem as the

plant matures. They are characterized by a rectangular shape, vacuoles in their protoplasm as well as tannins.

- Meristematic cells that do not differentiate into adult cells are known as initiating cells/meristematic initials.

General Characteristics of Meristem Cells

- Have a relatively thin, flexible cell wall (a thin cellulose wall) – Unlike in permanent, mature cells (differentiated cells), the cell wall of meristematic cells is thin and flexible. This allows the cell to continue actively dividing. For mature, differentiated cells, the rigid cell wall affects the ability to actively divide.
- Plastics in these cells are in the proplastids stage.
- Have less E.R compared to other cells.
- With regards to shape, they may appear rectangular, oval, spherical or polygonal.
- Lack intercellular space between them – Compact.
- Lack vacuoles - In the event that vacuoles are present, they are small in size.
- They are small and contain dense cytoplasm.

Meristem Cells Vs. Stem Cells

Essentially, stem cells are undifferentiated cells of multicellular organisms that give rise to both the same type of cells (as themselves) as well as other types of cells through differentiation.

As in animals, stem cells in plants are located in stem cell niches known as meristems (protomeristem and primary meristem). Here, plant cells that are regarded as stem cells are undifferentiated and located in the very tips of the plant.

Compared to stem cells in animals, studies have shown that stem cells in the meristem can be restored even in cases where they have been removed. In the event that favorable conditions are provided, this allows the plant to continue growing. Here, it is worth noting that plant stems are located in the meristem tissue at the growth points of the plant.

Adaptations of Meristem Cells

- Meristem cells can be restored: One of the most beneficial adaptations of is that they can be repeatedly restored. According to studies, meristem cells can arise from differentiated cells. This is made possible by the feedback principle between stem cells and actively dividing cells where a protein synthesized by the stem cells help restore meristem.

- Reprogramming: Types of meristematic cells can be reprogrammed to form different types of plant cells. In particular, this has been shown to be the case in tissue injury. Here, reprogramming allows specific types of cells to be produced to repair the injury.
- Small, polygonal or spherical in shape: This characteristic is important given that it allows for a large number of meristematic cells to be closely packed.

Parenchyma Cells

In plants, parenchyma is one of three types of ground tissue. Ground tissue is anything that is not vascular tissue or part of the dermis (skin) of the plant. In contrast to collenchyma and sclerenchyma cells, parenchyma cells primarily consists of all of the simple, thin walled, undifferentiated cells which form a large majority of many plant tissues.

Structure of Parenchyma Cells

Parenchyma cells are notable for their thin walls, and for being alive at maturity. Collenchyma cells tend to develop thicker secondary cell walls, to support structure. Sclerenchyma cells get both thicker walls and die off at maturity, producing tissues like bark and vascular tissue. The parenchyma cells have thinner walls and stay alive at maturity. While this makes them less useful in structural applications, the cells can move and store water and nutrients as well as divide quickly. This is important for the growth and repair functions of the parenchyma cells.

Each parenchyma cell may be a different shape, depending on its exact location and which tissue it is present in. However, it will always have a large central vacuole. This organelle is responsible for storing water and ions. This both creates a pressure between the parenchyma cells and their neighbours (called turgor pressure) and also allows the plant to store enormous amounts of water and nutrients. The thin walls of the parenchyma cells also allow the easy passage of sugars created in the leaves.

In fact, most photosynthesis takes place within specialized parenchyma cells found within leaves. These parenchyma cells, called chlorenchyma cells, contain chloroplasts. Chloroplasts are special organelles which carry out the process of photosynthesis, storing the energy of sunlight in the newly created bonds of sugar molecules. These sugars can then be converted into other sugars, fats, and oils, and stored in other parenchyma cells within the stems and roots. Potatoes, for example, are mostly parenchyma cells packed with stored starches. The plant would typically use the stores to survive the winter and get a boost the next spring.

Parenchyma Cells Functions

Healing and Repair

One of the most important functions of parenchyma cells is that of healing and repair. Parenchyma cells are unique in their meristematic nature. This means that the cells are pluripotent, having the ability to divide into a number of different cells. This plays an important role in how a plant can heal itself after a wound. While it may seem silly to think that a tree heals, the process is not much different to healing in a human body.

Parenchyma cells, once exposed to the outside when a wound occurs, are stimulated to start dividing. The cells divide towards the wound, differentiating into the different cell types which are needed, such as bark and epidermis. The parenchyma cells on the inside of the wound remain undifferentiated, and provide a source of meristematic cells in case the plant is attacked again. This process is responsible for healing in plants, from giant trees to a blade of grass.

Photosynthesis

Another important role parenchyma cells play is that of provider. While the other cell types provide much of the support and foundation on which the parenchyma cells operate, they produce a majority of the photosynthesis products. Simply through sheer numbers, parenchyma cells outnumber the other types. The chlorenchyma cells specifically do the majority of the photosynthesis.

However, photosynthesis would come to halt if the products had nowhere to go. Some parenchyma cells differentiate into part of the phloem, a special passageway for the sugars and products of photosynthesis to traverse the plant. These parenchyma cells allow the products to make it from the leaves, where they are created, all the way to the roots. The living cells have specialized proteins and channels which are used to help the sugars make their way efficiently to the roots and other tissues. These other parenchyma tissues need the sugars because they are internal and do not contain chloroplasts with which to create their own energy.

Nutrient and Food Storage

Humans rely on the storage ability of parenchyma cells as our main source of food. The entire food chain is based upon the storage of sugar within parenchyma cells. So, whether you eat meat or are a vegan, you need parenchyma cells. The large central vacuole within plant cells allows the storage of large amounts of soluble nutrients, which dissolve into the water. The plant can control the usage and distribution of the nutrients within cells via the activation of specific proteins and pathways. Parenchyma cells are a major storage place for ions, water, and all photosynthesis products. Many of the foods we know, like fruits and vegetables, are purposefully bred exaggerations of natural plant processes. Corn, potatoes, and wheat were all selected from less productive ancestors which stored higher amounts of nutrients in their parenchyma cells.

Collenchyma Cells

In plants, Collenchyma is a cell and tissue type which supports tissue of living elongated cells with irregular cell walls. Collenchyma cells have thick deposits of cellulose in their cell walls and appear polygonal in cross section. The strength of the tissue results from these thickened cell walls and the longitudinal interlocking of the cells. Collenchyma may form cylinders or occur as discrete strands and is one of the three ground, or fundamental, tissues in plants, together with parenchyma (living thin-walled tissue) and sclerenchyma (dead support tissue with thick cell walls).

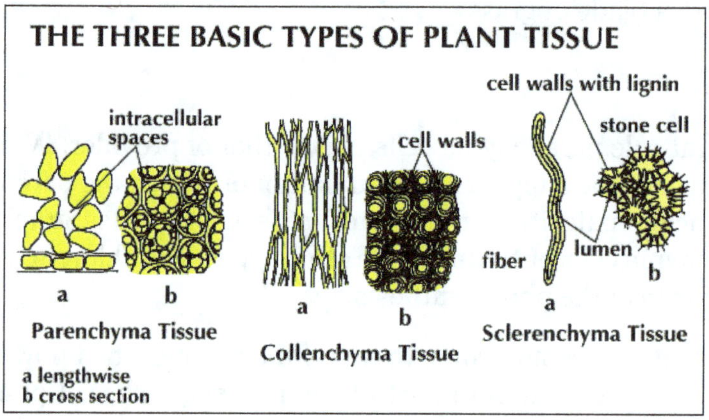

Ground tissue.

Three types of ground, or fundamental, tissue in plants. Parenchyma tissue is composed of thin-walled cells and makes up the photosynthetic tissue in leaves, the pulp of fruits, and the endosperm of many seeds. Collenchyma cells mainly form supporting tissue and have irregular cell walls. They are found mainly in the cortex of stems and in leaves. The major function of sclerenchyma is support. Unlike collenchyma, mature cells of this tissue are generally dead and have thick walls containing lignin. Their size, shape, and structure vary greatly.

An important feature of collenchyma is that it is extremely plastic—the cells can extend and thus adjust to increased growth of the organ. The tissue is found chiefly in the cortex of stems and in leaves and is the primary supporting tissue for many herbaceous plants. In plants with secondary growth, the collenchyma tissue is only temporarily functional and becomes crushed as woody tissue develops. It often constitutes the ridges and angles of stems and commonly borders the veins in eudicot leaves. The "strings" in stalks of celery are a notable example of collenchyma tissue.

Close-up of veins in a leaf.

Distribution

Collenchyma cells are present at the periphery of herbaceous stems, petioles (e.g. Begonia) and in the ribs of some leave (e.g. Nerium). They are also present in the floral parts, fruit and aerial root (ex. Monstcra). They usually occupy the peripheral layers of cortex in dicotyledons and may be present just beneath the epidermis or below a few peripheral layer of parenchyma.

They may be present as a continuous layer to form hypodermis. Sometimes the cells form distinct ribs or ridges at the outer edges of stems (e.g. Leonurus, Cucurbita etc.) and petioles (e.g. Apium). They are uncommon in roots, monocotyledonous leaves and stems. Subterranean roots of Vitis and Diapensia contain collenchyma. The rind of fruits is collenchyatous in Vitis and Cassia tora.

Origin

Collenchyma develops from the ground meristem or from procambium (e.g. Apium graveolens). They may also originate from elongated cells, which resemble procambium.

Shape

Isolated collenchyma ranges from round, elliptical to elongated cell Transitional forms are also present. The round or elliptical cells resemble ground parenchyma. The cells may assume the shape of a short prism. The elongated cells look like fibre with both ends tapered and may attain a length of 2 mm though rarely.

Arrangement

Collenchyma tissue appears to be more or less compactly arranged as the thickening materials deposit more heavily at the corners and on the radial walls of cell in addition to normal uniform thickening.

The nature of wall thickenings varies and accordingly the following three types of collenchyma are distinguished:

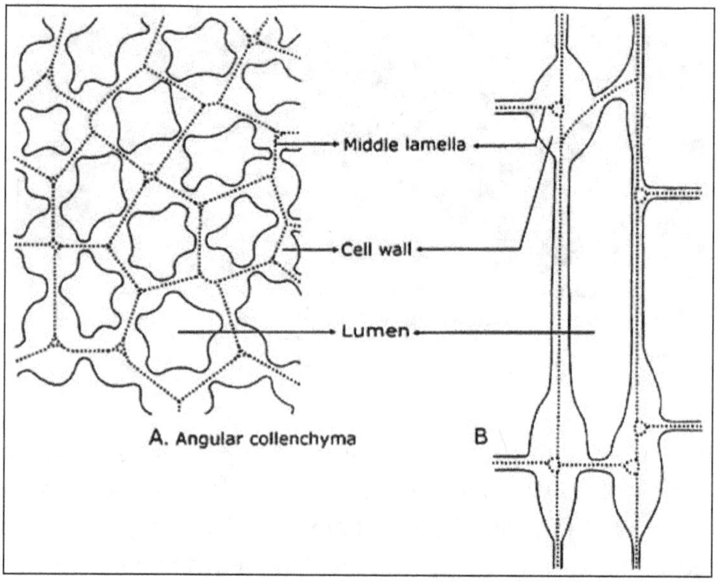

A: Diagram of collenchyma cells in cross sectional view.
B: Line drawing of angular collenchyma cells in logitudinal section.

Types of Collenchyma

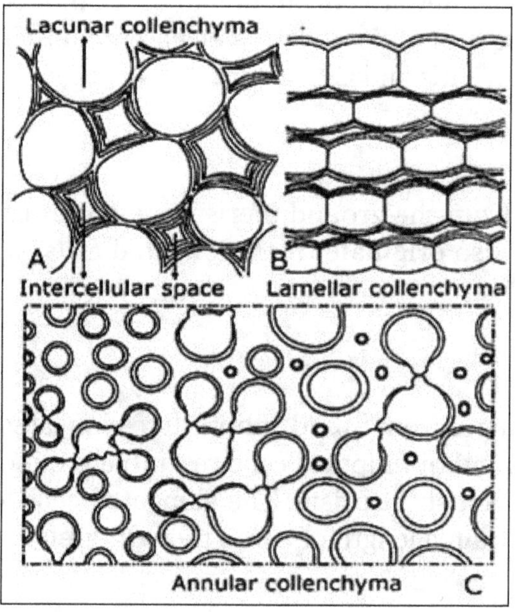

Collenchyma cells in cross-sectional view.

- Angular: The thickening materials deposit at the angles or corners of the cells. In cross sectional view the thickenings occur at those places where several cells meet. Example: petioles of Cucurbita, Beta etc.

- Lamellar: The thickening materials deposit heavily on the tangential walls of the cell than the radial walls, ex. stem of Sambucus.
- Lacunar: The thickening materials deposit at those places of the cell wall, which are in direct contact with the intercellular spaces. Example: petioles of Salvia, Malva etc.

Researcher reported another type, called annular collenchyma in the petiole of Nerium. In this type the cell lumen appears to be more or less circular in cross sectional view. Fahn opined that this might be the case of mature collenchyma. It is observed that due to continued and heavy deposition of wall materials the angular appearance of the lumen may be lost.

Structure and Content

The walls are thickened by high amounts of pectin and hemicellulose. Protein and cellulose are also present. Primary pit fields can be distinguished in the walls. The cellulose microfibrils are either oriented transversely or alternately transverse and longitudinal. The longitudinal microfibrils may also alternate with noncellulosic materials, as is revealed by electron microscopic study with the petiole of Apium graveolens.

The composition of these alternating layers varies. One layer is rich in cellulose and poor in pectin; the alternate layer is pectin-rich and cellulose-poor. It is now suggested that the visible lamellation of the cell wall of collenchyma is due to composition and orientation of microfibrils at alternate layers. Usually lignin is completely absent.

But in some cases the cellulose-rich layer may be impregnated with lignin (e.g. stem and petioles of Salvia officinalis, Viscum album, Medicago sativa etc.). Lignified lamellae may also be deposited on the wall. Thus the pecto-cellulosic wall of collenchyma may become sclerified. Collenchyma contains living protoplast. Chloroplastids may also be present in some cells. The cells may also contain tannins.

Function

- The cells are extensible with a considerable degree of plasticity and so support the organs in which they occur.
- Chloroplast containing collenchyma can carry out photosynthesis.
- In some cases the peripheral thick walled collenchyma becomes thin and regains the meristematic activity, e.g., phellogen, the cork cambium, which divides to form the periderm.
- The sclerified collenchyma is the mechanical cell of mature plants.
- Collenchyma can serve as storage sites for defensive antibacterial compounds.

The compounds are hydroxyproline-rich bacterial agglutinins found in Solanum tuberosum.

Sclerenchyma Cells

In plants, Sclerenchyma is a cell and tissue type which supports tissue composed of any of various kinds of hard woody cells. Mature sclerenchyma cells are usually dead cells that have heavily thickened secondary walls containing lignin. The cells are rigid and no stretchable and are usually found in no growing regions of plant bodies, such as the bark or mature stems. Sclerenchyma is one of the three types of ground, or fundamental, tissue in plants; the other two types are parenchyma (living thin-walled tissue) and collenchyma (living support tissue with irregular walls). Sclerenchyma cells occur in many different shapes and sizes, but two main types occur: fibres and sclereids.

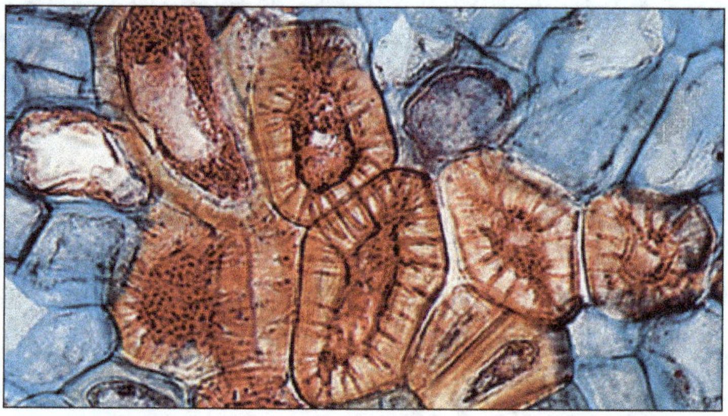

Sclerenchyma: Cross-section of sclerenchyma tissue (red). The thickened, secondary walls are lignified to provide support to the surrounding plant tissues.

A worker stripping fibre from abaca (*Musa textilis*) in the Philippines.

Fibres are greatly elongated cells whose long, tapering ends interlock, thus providing maximum support to a plant. They often occur in bundles or strands and can be found almost anywhere in the plant body, including the stem, the roots, and the vascular bundles in leaves. Many of these fibres, including seed hairs, leaf fibres, and bast fibres, are important sources of raw material for textiles and other woven goods.

Sclereids are extremely variable in shape and are present in various tissues of the plant, such as the periderm, cortex, pith, xylem, and phloem. They also occur in leaves and fruits and constitute the hard shell of nuts and the outer hard coat of many seeds. Sometimes known as stone cells, sclereids are also responsible for the gritty texture of pears and guavas.

Nut: The hard shells of many nuts contain sclereids, which are a type of sclerenchyma cell.

Characteristics of Sclerenchyma Tissue

Term sclerenchyma was derived from the Greek word "Scleros" that means harder and "Enchyma" which means infusion. A sclerenchyma tissue shows the following characteristic features:

- It is a dead, simple-permanent ground tissue.
- The function of sclerenchyma is similar to the collenchyma tissue, which is giving mechanical support and tensile strength to the plants.
- Sclerenchyma cells function as a "Skeleton" of the plant system that contributes rigidity to withstand against various ecological stresses.
- During the initial growth cycle of a plant, the sclerenchyma persists as a living cell and found as a spiral or ring patterns.
- On plant maturation, the sclerenchymatous cells become dead by the accumulation of lignin that makes the cell harder and impervious to the exchange of water, solutes, gases etc. between the environment and the inner protoplast.

- Sclerenchyma refers as a dead tissue because of its dead, degenerated or functionless inner protoplast.
- Mechanical and conductive sclerenchymatous tissues are two common types, based on the function of sclerenchyma.
- Fibres, sclereids and treachery elements are the three common types, based on the morphology of sclerenchyma tissue.
- It includes two layers of cell-wall; a primary cell-wall and a thickened secondary cell-wall (containing cellulose, hemicellulose, lignin etc.).
- The cell-wall type, rigidity, shape, size etc. of sclerenchyma will vary accordingly, within different types of plant.

Types of Sclerenchyma Tissue

Based on the function: A sclerenchyma tissue can classify broadly into two classes, namely mechanical and conductive sclerenchyma.

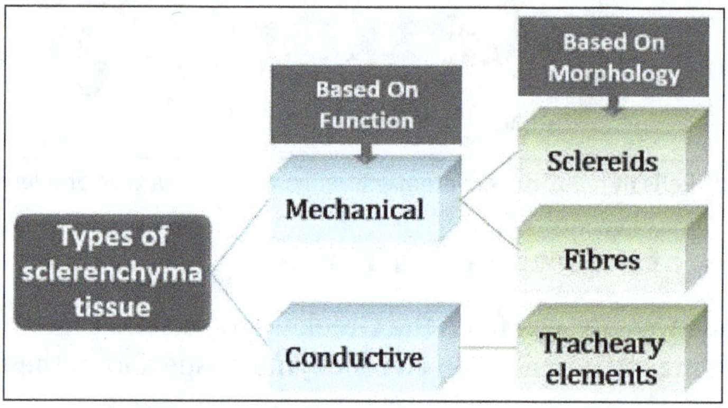

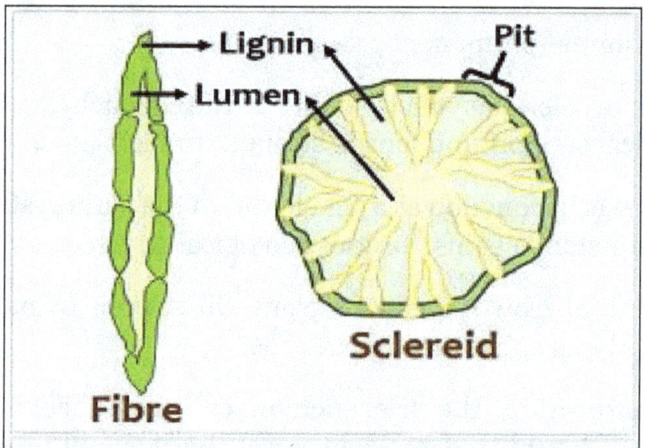

SCLERENCHYMATOUS TISSUE

Mechanical Sclerenchyma

It is a kind of sclerenchymatous tissue that functions as a "Supportive tissue" by reducing the wilting in plants, maintaining plant physiology, providing strength to withstand against the tearing forces of waves and current etc. Mechanical sclerenchyma comprises of sclereid and fibre cells that contribute strength and stiffness to the plant system.

Sclereids

Sclereids can define as a mechanical tissue that involves the following features:

- It occurs singly or in groups.
- Sclereids are found associated with the plant's vascular tissue, namely xylem and phloem.
- Its cell-wall thickening is non-uniform and contains a number of simple pits with round apertures.
- Sclereids most usually comprises of the narrow lumen.

Based on the shape, the sclereids cells subdivides into the following classes:

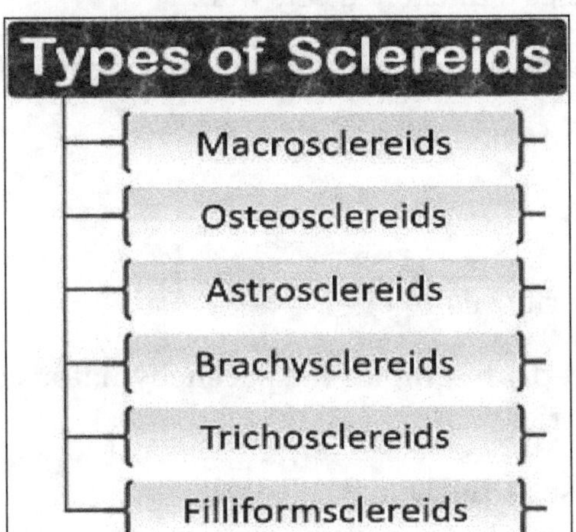

Macrosclereids

- It also refers as "Malpigian cell".
- Appearance: Elongated and columnar in shape.
- Cell wall: Comprises of a thickened cell-wall.
- Occurrence: Usually found in the outer epidermal cells of seed.
- Example: Seed coat of Pisum species.

Osteosclereids

- It also refers to as "Bone cells".
- Appearance: Appears very similar to the shape of a bone of hourglass with enlarged, lobed and columnar cells. It is lobed towards the end.
- Cell wall: Comprises a thickened cell-wall.
- Occurrence: Usually found below the epidermal layer i.e. hypodermis of seeds and leaves of certain plants belongs to the category of Xerophytes.
- Example: Leaves of Hakea species.

Astrosclereids

- It also refers to as "Stellate cells".
- Appearance: These appear to be star-like, deeply lobed with the radiating arms from the central body. The radiating arms are usually pointed, irregular and varied in number.
- Cell wall: Comprises a thickened cell-wall.
- Occurrence: Extends from upper to lower epidermis of the leaf.
- Example: Leaves of Thea, Olea etc.

Brachysclereids

- It also refers to "Grit cells".
- Appearance: It deeply resembles the parenchymatous cells, and its symmetry is roughly isodiametric.
- Cell wall: Comprises a thickened cell-wall.
- Occurrence: It is commonly present in the fleshy portions of fruit.
- Example: Flesh of pear fruit, where brachysclereids form a grit and also refers as stone cells.

Trichosclereids

- It also refers to as "Needle-like cells".
- Appearance: It seems to be hair-like, more elongated, and branched cells stretching towards the intercellular space.

- Cell wall: Comprises a thickened cell-wall.
- Occurrence: Present in the specialized tissues of leaves and roots.
- Example: Aerial roots of Monstera sp, leaves of olive and water-lily etc.

Filiform Sclereids

- It also refers to as "Fibre-like cells".
- Appearance: These are very much elongated, sparingly-branched and uncommon kind of a cell.
- Cell wall: Comprises a thickened cell-wall.
- Occurrence: Found in the specialized tissues of leaves.
- Example: Leaves of Olea.

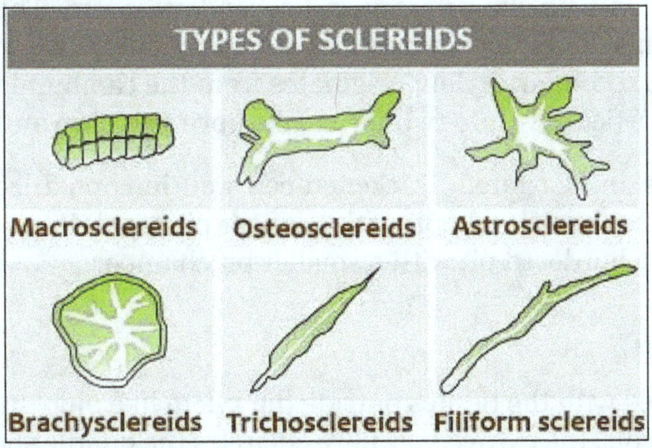

Functions of Sclereid

- Sclereids support the neighbouring tissues where they occur.
- It protects the inner cells, by forming a concentrating layer towards the periphery.

Fibres

Fibres can define as another kind of mechanical tissue that involves the following features:

- The fibre cells are elongated, thick-walled with a narrow lumen and tapered ends.
- It occurs in the ground and vascular tissues of a plant.

- Fibre sclerenchyma is a cell companion to the xylem and phloem.
- It appears as independent strands or cylinders.

Based on the shape, the fibre cells subdivide into the two following groups:

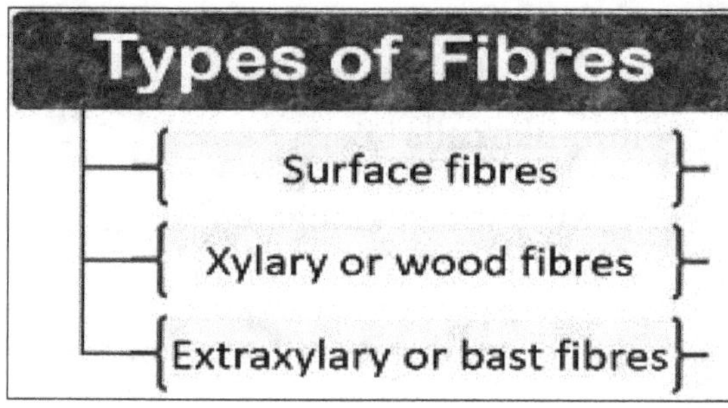

Xylary Fibre

It is found associated with the primary and secondary xylem. The xylary fibre associated with the primary xylem originates from the procambium, whereas xylary fibre associated with the secondary xylem originates from the cambium tissue of plant cell. It further divides particularly into two forms like libriform fibre and fibre tracheids.

Libriform fibre has an elongated, thickened cell wall in comparison to the fibre tracheids, and comprises of a simple pit with a longer pit canal. Fibre tracheids are long, thick-walled, having bordered pit with a smaller pit chamber.

Extraxylary Fibre

It is found associated with the tissues outside the xylem like phloem, cortex and pith of plant cell. In monocots, the extraxylary fibre encircles the bundle sheath, derives partly from the ground meristem and remaining from the procambium.

In dicots, the extraxylary fibres occur as independent bands or cylinders, on the peripheral region of the vascular cylinder and innermost cortex layer. It derives wholly from the ground meristem tissue, whose structure; shape and composition are somewhat similar to the xylary fibre.

The extraxylary fibre subdivides into three kinds like:

- Phloem fibres: These occur in the primary and secondary phloem of vascular plant tissues, and also called as “Bast fibres”.
- Cortical fibres: These are present in the cortex region of a plant cell that occurs singly or in groups, and supports the younger part of plants.
- Perivascular fibre: It is present in the pericycle of the plant, forming a vascular bundle cap of dicot and bundle sheath of monocots, and also refers as “Pericyclic fibres”.

Functions of Fibre Tissue

- Fibre tissue contributes flexibility to the plant.
- The septate fibres function as storage cells that reserve starch and oil droplets.
- It protects the nearby inner tissue.
- The surface fibres facilitate seed and fruit dispersal.
- Plant fibres help in the manufacturing of textile, ropes, strings etc.

Conductive Sclerenchyma

It consists of a tracheary element that is a peculiar property of vascular plants, which demarcates them from the non-vascular plants. The tracheary elements provide both strength and water conduction.

Tracheary Elements

These elements occur in the vascular plants that include vessel elements and tracheids.

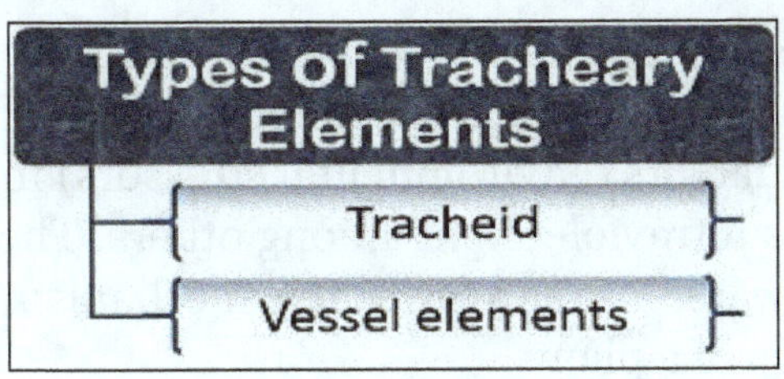

Vessel Elements

It comprises of a perforated end walls (primary and secondary lignified wall) and present in both primary and secondary xylem. Vessel elements are more efficient in the conduction of water, where the water flows vertically from one cell to the other without any hindrances.

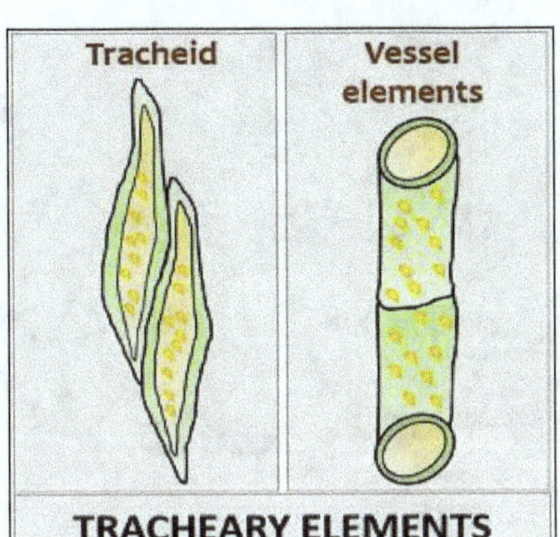

TRACHEARY ELEMENTS

It is a more specialized kind of tracheary element and the size is smaller than the tracheids. The vessel elements interconnect with the other vessels from one end of the cell to the cell of another end, in vertical rows. One of the characteristic features of vessel element is that it solely participates in the conduction of water.

Tracheids

Tracheids are the common cell in the xylem that appears to be spindle-shaped, elongated with tapered ends. It participates in both water conduction and mechanical support. Tracheids are elongated comparative to the vessel elements, and having a common feature of having secondary wall thickening ranging into various shapes (from annular rings, reticulate etc. to pitted form). It is having a high surface to volume ratio that is one of the characteristic property of tracheids, which protects the plant from air embolisms or water stresses.

Epidermal Cells

Epidermal cells include several types of cells that make up the epidermis of plants. Although they serve a number of important functions, their primary role is to protect from a variety of harmful factors (environmental stressors) including microbes, chemical compounds as well as ultraviolet light among others. These cells are situated very close together to prevent water loss as a protective mechanism. The cell layer covers the seeds, stem, root and leave of a plant.

In plants:

- Pavement cells
- Stomatal guard cells
- Trichome

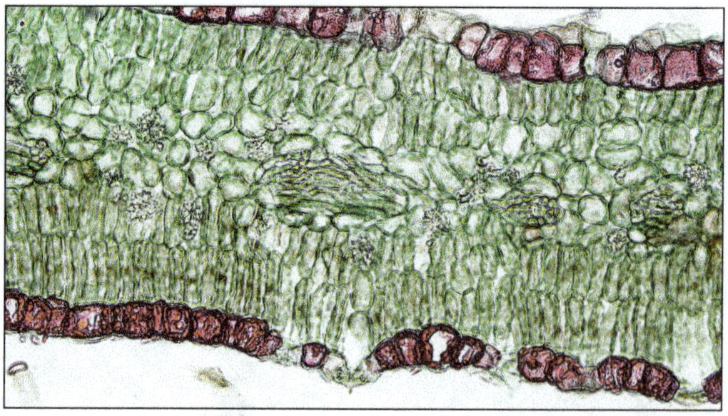

Structure and Functions of Epidermal Cells

Plant Epidermis

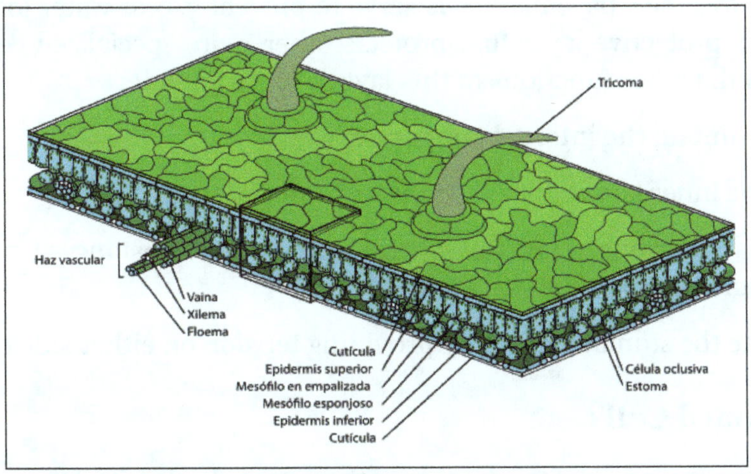

In plants, differentiation of the epidermal cells occurs during embryogenesis in a developing seed. Like the skin epidermis, the epidermis of the plant covers the outer surface and thus covers all plant tissue from the roots to the tip. Made up of epidermal cells, the epidermis in plants also serves as a protective layer that not only prevents various microorganisms from gaining entrance into the underlying tissue of leaves and stems, but also prevents excess water loss among a few other functions. Like the skin epidermis, epidermis of plants also consists of different types of cells that vary in morphology and serve different function.

Plant Epidermal Cells

Pavement Cells

Pavement cells are the most common cells of the plant's epidermis. As such, they can be found covering all plant organs in any plant. As compared to the other types of cells, pavement cells are not fully specialized. For this reason, their shapes (morphology) are not well modified for special functions as is the case with stomatal guard cells. For different plants and organs, however, studies have shown the morphology of pavement cells to vary.

For instance, in Arabidopsis thaliana, pavement cells have an irregular wavy shape that is produced during the development of leaves. In the leaves of many dicots, the shape resembles interlocking jigsaw puzzle pieces which provide some mechanical strength to the leaves.

As compared to other parts of the plant, pavement cells located in the stem and various elongated plant organs have a rectangular appearance with a long axis that is parallel to the direction of expansion (of the organ/stem). The differences in morphology have been attributed to the functions and growth forms of these organs. Epidermal of pavement cells in coma plants (Arabidopsis) have been shown to contain chloroplasts.

Tightly packed together, pavement cells serve to prevent excess water loss. In addition, they make up a protective layer that protects other more specialized cells located beneath. Some of the other functions of this layer of cells include:

- Help maintain the internal temperature.
- Keep the inner layers of cells in place.
- Barriers to various organisms, particles and other substances from the external environment.
- Separate the stomata apart (by providing tension on either side of stomata).

Stomatal Guard Cells

Stomatal guard cells are part of the epidermal tissue that serves several functions in plants. Depending on the type of plant, the spatial arrangement of these cells is not only dependent on size, but also the shape of air-space below them. Unlike pavement cells, guard cells are more specialized with a definitive shape that allows them to carry out their functions.

With regards to structure, two guard cells form the stomata. Depending on water availability (as well as the concentration of sugars and ions), guard cells can become turgid which controls the closing and opening of the stomata pore. In turn, the closing and opening of these pores regulate gaseous exchange in and out of the leaves:

- Turgor pressure regulates the closing and opening of guard cells.
- Guard cells also contain chloroplasts that allow for photosynthesis.

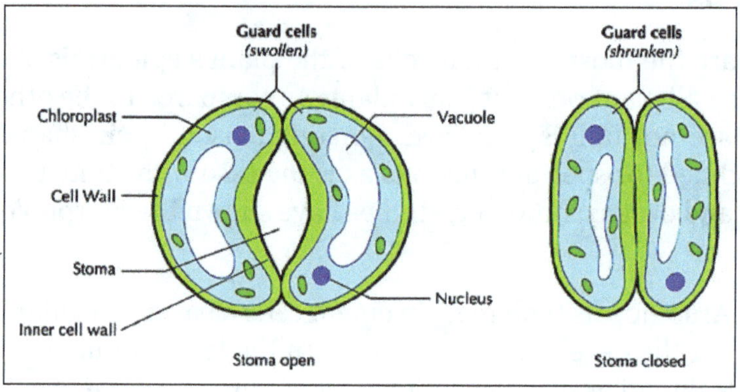

Trichomes

Trichomes (epidermal hairs) are tiny hairs located on the epidermal tissue. Like stomatal guard cells, trichomes are also more specialized and thus have well-defined shapes that contribute to their functions. The trichome of Arabidopsis has been well studied and described over the years.

With large single cells measuring between 200 and 300um in length, different types of trichome have been shown to play a protective role in plants where they protect plants from predators as well as organisms that cause diseases.

Here, the trichome achieves this by either trapping or poisoning the animal to protect the plant. For some of the plants, however, trichomes simply function as barriers that protect inner tissues of leaves.

Unlike the other cells of the epidermal tissue, studies have shown that cell division is arrested in trichomes. Several rounds of endoreduplication are therefore responsible for the expansion of the cell as pavement cells continue dividing.

Cuticle

In plants leaves, epidermal cells are located on the upper and lower part of the leaf where they form the upper and lower epidermis. The cuticle, however, is located on the upper epidermis for the most part. In plants, this is the outermost part that is secreted by the epidermis. Here, it consists of a substance known as the cutin (polymerized esters of fatty acids). On the upper epidermis, the cuticle, which is waxy in nature, acts as a water-repellent? It is also shiny and thus helps reflect off excess sunlight. The thickness of a cuticle in plants is largely dependent on the type of plant and where they are located.

Apart from plants, the cuticle can also be found in various organisms such as arthropods. Here, it consists of various pigments and chitin that protect the inner tissue of the organism. In human beings, however, the cuticle is the epidermis.

Epidermal Cells in Onion

Epidermal cells of onions are very simple. As a result, the epidermal tissue has become the ideal model through which students are introduced to the morphology/anatomy of plant cells.Epidermal cells of onions also have well-defined shapes that may appear rectangular or square (or as elongated hexagonal) under the microscope.

When viewed under the microscope, it is possible to view the cell nucleus, a very thin layer of cytoplasm that can be seen in some of the cells as well as the cell walls at the boundary of each cell. These are characteristics of living cells that are capable of division and growth. Unlike epidermal cells of various plants, epidermal cells of onions have a layer of one cell in thickness. Some of the other components of epidermal cells of onions include:

- Actin microfilaments.

- A middle lamina that contains pectin.
- Vacuole.

Like other epidermal cells, the primary function of epidermal cells of an onion is to protect underlying tissue against such microorganisms as viruses.

Xylem Cells

Xylem is a plant vascular tissue that conveys water and dissolved minerals from the roots to the rest of the plant and also provides physical support. Xylem tissue consists of a variety of specialized, water-conducting cells known as tracheary elements. Together with phloem (tissue that conducts sugars from the leaves to the rest of the plant), xylem is found in all vascular plants, including the seedless club mosses, ferns, horsetails, as well as all angiosperms (flowering plants) and gymnosperms (plants with seeds unenclosed in an ovary).

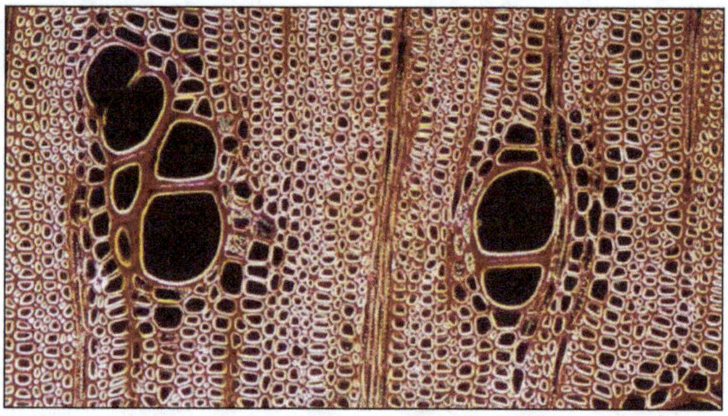

Xylem tissue: Cross section of xylem tissue from an oak tree (*Quercus* species). The large holes are wide vessel members, which serve as major water-conducting cells for the tree.

The xylem tracheary elements consist of cells known as tracheids and vessel members, both of which are typically narrow, hollow, and elongated. Tracheids are less specialized than the vessel members and are the only type of water-conducting cells in most gymnosperms and seedless vascular plants. Water moving from tracheid to tracheid must pass through a thin modified primary cell wall known as the pit membrane, which serves to prevent the passage of damaging air bubbles. Vessel members are the principal water-conducting cells in angiosperms (though most species also have tracheids) and are characterized by areas that lack both primary and secondary cell walls, known as perforations. Water flows relatively unimpeded from vessel to vessel through these perforations, though fractures and disruptions from air bubbles are also more likely. In addition to the tracheary elements, xylem tissue also features fibre cells for support and parenchyma (thin-walled, unspecialized cells) for the storage of various substances.

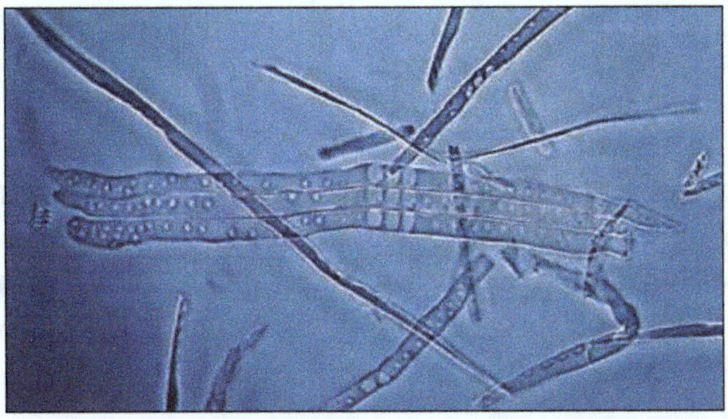

Xylem tracheids: Tracheid plant cells: As part of the xylem tissue, tracheids conduct water and minerals from the roots to the rest of the plant.

Xylem formation begins when the actively dividing cells of growing root and shoot tips (apical meristems) give rise to primary xylem. In woody plants, secondary xylem constitutes the major part of a mature stem or root and is formed as the plant expands in girth and builds a ring of new xylem around the original primary xylem tissues. When this happens, the primary xylem cells die and lose their conducting function, forming a hard skeleton that serves only to support the plant. Thus, in the trunk and older branches of a large tree, only the outer secondary xylem (sapwood) serves in water conduction, while the inner part (heartwood) is composed of dead but structurally strong primary xylem. In temperate or cold climates, the age of a tree may be determined by counting the number of annual xylem rings formed at the base of the trunk (cut in cross section).

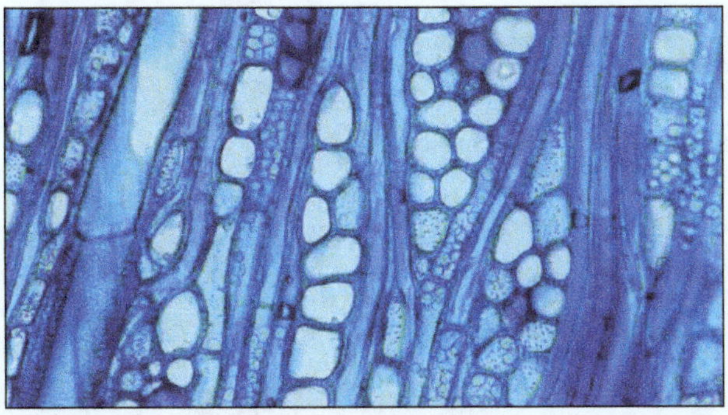

Root xylem: Stained and magnified root tissues of Texas persimmon (*Diospyros texana*) from Colorado Bend State Park, Texas. The stacks of circular cells are the water-conducting xylem tissues of the root.

Function of Xylem

The main function of xylem is to transport water, and some soluble nutrients including minerals and inorganic ions, upwards from the roots to the rest of the plant. Xylem cells form long tubes that transport materials, and the mixture of water and nutrients that flows through the xylem cells is called xylem sap. These substances are transported through passive transport, so the process doesn't require energy. The phenomenon that allows xylem sap to flow upwards against gravity is called capillary action. This occurs when surface tension makes liquid move upward. Water is also aided in moving up through the xylem by adhering to the xylem cells. However, it gets harder to work against gravity to transport materials as a plant grows taller, so xylem sets an upper limit on the growth of tall trees.

Xylem evolved in plants over 400 million years ago. In order to make food through photosynthesis, plants need to absorb carbon dioxide from the atmosphere and water from the soil. However, when the stomata—small holes in a plant's leaves—are open to allow CO_2 in, a lot of water evaporates, much more than the amount of CO_2 taken in. Plants that developed systems to transport water to the sites of photosynthesis on leaves had a better chance of survival.

Structure of Xylem

Xylem is made up of several types of cells. Tracheids are long cells that help transport xylem sap and also provide structural support. Vessel elements are shorter than tracheids, but also help conduct water. They are found in flowering plants, but not in gymnosperms like pine trees. Vessel elements have perforation plates that connect each vessel element to form one continuous vessel. Xylem also contains parenchyma, a tissue that makes up most of the soft parts of plants, and long fibers that help support the plant. In a cross section of a plant, under a microscope, xylem appears star-shaped.

This figure describes different parts of the xylem:

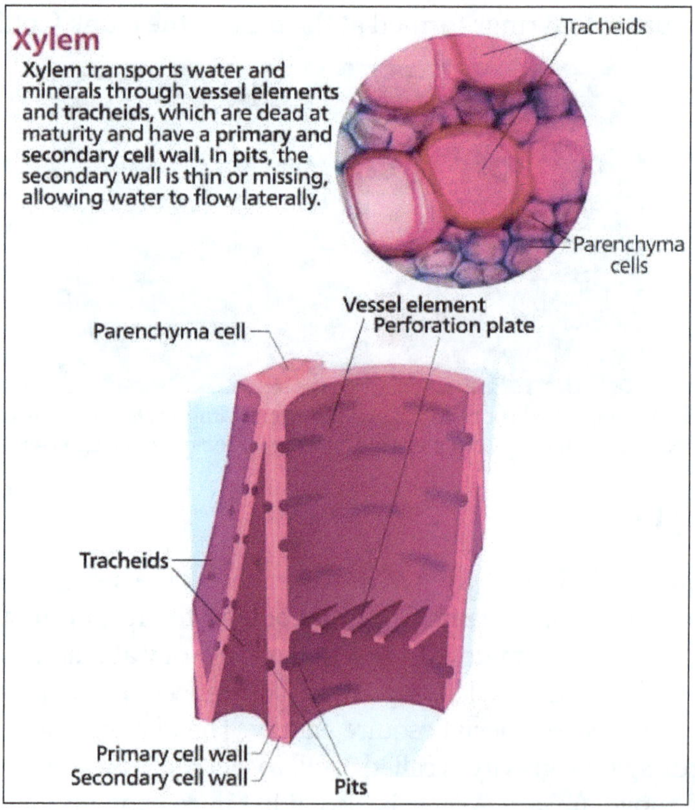

Details of Xylem Development

The first xylem that develops in a growing plant is called protoxylem, and it contains narrow vessels as the plant is not yet big. Metaxylem develops later on and has larger vessels and cells. There are four ways protoxylem and metaxylem can be arranged in a plant: centrarch, exarch, endarch, and mesarch. Centrarch: Xylem forms one chamber in the middle of the stem, with metaxylem surrounding the protoxylem. This formation is not found in any living plants today:

- Exarch: Xylem develops in multiple strands, and each strand develops inward towards the center of the root. Xylem in roots of vascular plants develops this way.
- Endarch: Xylem develops in multiple strands, and each strand develops outward towards the periphery of the stem. Xylem in stems of vascular plants develops this way.
- Mesarch: Xylem develops in multiple strands, and each strand develops from its middle both toward the center of the stem and in the opposite direction toward the periphery. Xylem in leaves and stems of ferns develop this way.

Types of Xylem

The two types of xylem, primary and secondary, perform the same function but are categorized by the type of growth that they are formed with:

- Primary Xylem: Primary xylem forms with primary growth of a plant. This is the growth that occurs at the tips of stems, roots, and flower buds. It allows the plant to grow taller and the roots to grow longer. This growth is called primary because it occurs first in the growing season, before secondary growth. Both primary and secondary xylems transport water and nutrients.
- Secondary Xylem: Secondary xylem is formed with a plant's secondary growth; this is the type of growth that allows the plant to get wider over time. Wide tree trunks, for example, show a lot of secondary growth. It occurs each year after primary growth. Secondary xylem is what gives the inside of tree trunks dark rings that are used to determine the age of the tree.

Differences between Xylem and Phloem

Xylem and phloem both make up the vascular system of the plant, and work together to form vascular bundles that provide mechanical strength to the plant, but they have important differences. While xylem transports water, phloem transports food and nutrients. (One way to remember this is that phloem and food both begin with an "F" sound.) Xylem is unidirectional; its job is to make sure water flows upward. However, phloem is bidirectional and transports food and nutrients to all of the plant. Mature xylem is made up of dead cells that do not have cell contents, while phloem contains living cells (albeit without nuclei). The structure of xylem and phloem is also different. While xylem is made up of tracheids and vessels, phloem is made up of sieve tubes which have many holes for transporting nutrients. Xylem is star-shaped, while phloem is round and actually surrounds the xylem.

Phloem Cells

Phloem is the vascular plant tissue responsible for the transport and distribution of sugars produced by the photosynthesis. Since the plant is a continuum, phloem will be found in the external part of root cylinders, in the stem vascular bundles and in the abaxial part of the venations of every single leaf. While the most common is to have the phloem external to the xylem in roots and stems and abaxial in leaves, some exceptions exist and are usually taxon specific. The phloem found in the inside is named internal or intraxylary phloem.

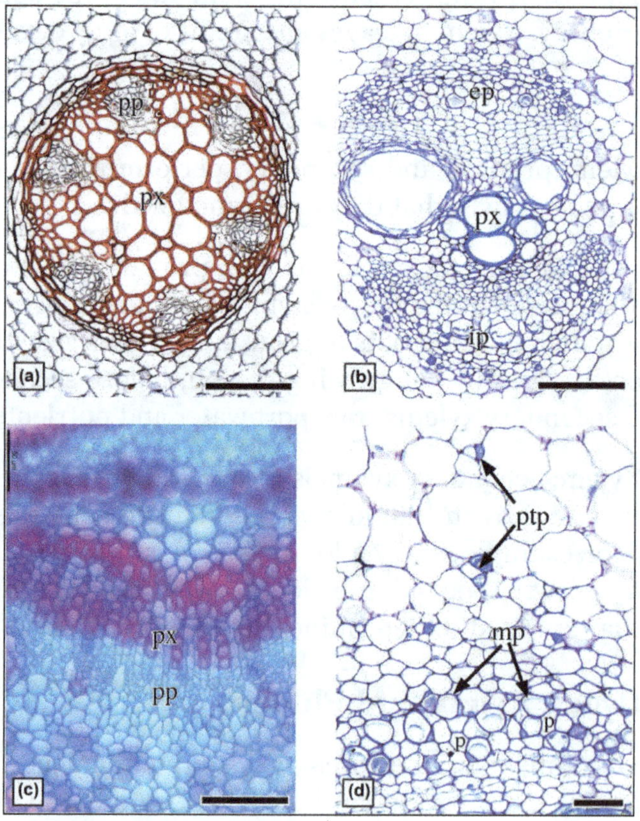

Figure shows Location of the primary phloem in different organs and its cell composition. (a) Ranunculus acris (Ranunculaceae). Root transverse section (TS), exarch structure, six strands of primary phloem alternating with the six protoxylem poles.(b) Bicollateral vascular bundle of a squash, Cucurbita pepo (Cucurbitaceae) TS. On the top is the external phloem, and on the bottom is the intraxylary or internal phloem. (c) Detail of the leaf midrib vascular cylinder of Tetrapterys mucronata (Malpighiaceae) showing primary xylem on the top and primary phloem on the bottom. (d) Detail of (b), showing the protophloem on top and the metaxylem on the bottom. ep, external phloem; ip, intraxylary phloem; mp, metaphloem; p, parenchyma cell; pp, primary phloem; ptp, protophloem; px, primary xylem. Scalebars: a, c, d = 50 μm, b = 130 μm.

As a constitutive tissue in the plant body, phloem functions extrapolate its main function of sugar transport, including transport of signalizing molecules such as mRNAs, hormones, defenses from biotic and abiotic agents, sustenance of the organs, gas exchange, and storage of many ergastic materials, such as starch, calcium oxalate crystals, and tannins. Parenchymatic cells of the phloem can also give rise to new meristems, such as the phellogen or cork cambium. All vascular plants have phloem, which typically includes specialized living conducting cells named sieve elements whose nucleus, ribosomes, and other organelles degenerate during maturation, making sugar transport more efficient. The life and function of these cells will then rely on closely associated parenchyma cells which support the physiological functions of these sieve elements. Although typical phloem is exclusive of vascular plants, rudimentary phloem-like conducting cells are present also in other lineages, such as the bryophyte leptoids, and even outside the plant kingdom, as the trumpet cells of the kelps and phaeophycean algae. The primary phloem derives from the embryo and the apical meristem procambium throughout the life of the plant or from the cambium, in plants with secondary growth.

Phloem Cell Types

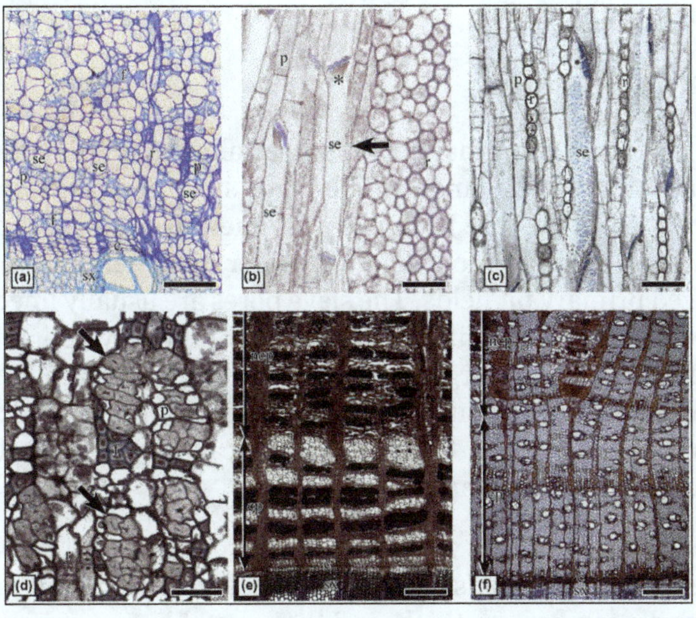

Figure shows General aspects of the secondary phloem. (a) Composition of the secondary

secondary phloem of Luehea divaricata (Malvaceae) TS, showing sieve tube elements (se) in clusters, axial parenchyma cells (p), fiber clusters (f), and rays (r). (b) Longitudinal tangential section (LT) of Cordia caffra (Boraginaceae) showing sieve tube element (se), companion cells (arrow), multiseriate ray (r), and axial parenchyma (p). Note callose staining with resorcin blue evidencing the slightly inclined simple sieve plates. Note also the P-protein (asterisk) next to the sieve plate. (c) LT of the secondary phloem of Castanea dentata (Fagaceae) showing sieve tube elements (se) with inclined, compound sieve plates and numerous lateral sieve areas of narrower pores, unicellular rays (r), and axial parenchyma (p). (d) TS of Talauma sp. (Magno-liaceae) showing sieve tube elements in clusters, with conspicuous nacreous walls, parenchyma cells (p), clusters of fibers (f), and rays (r). (e) Secondary phloem of maple, Acer saccharum (Sapindaceae), showing the conducting phloem (cp), where sieve tubes and companion cells are turgid, and the nonconducting phloem (ncp), with collapsed sieve tubes. (f) Secondary phloem of Carya cordiformis (Juglandaceae) showing a phloem formed by a background of fibers where solitary to multiple of two sieve tubes are scattered, with sieve-tube-centric and diffuse-in-aggregate axial parenchyma. Note that no collapse is seen in the nonconducting phloem of Carya. c, cambium; sx, secondary xylem. Scalebars: a = 100 μm, b-d = 50 μm, e, f = 200 μm.

The phloem is a complex tissue and is formed typically by three cell types, the sieve elements, the parenchyma cells, and the sclerenchyma cells. Sclerenchyma cells might sometimes be absent in primary and/or secondary phloem. The presence, quantities, and arrangements of these cell types in the tissue commonly vary and may be taxonomic informative. Lists depicting these variations in all phloem cell types are of ultimate importance for complete bark descriptions. What follows is a description of these three major cell types in the phloem.

Conducting Phloem Cells

Sieve element is a general term that encompasses all conducting cells of the phloem, both sieve cells and sieve tube elements. The name sieve derives from the strainer appearance given to the cells by the presence of numerous pores crossing their bodies. These pores are specialized plasmodesmata of wider diameter, and the sieve areas are basically specialized primary pit fields. The sieve pores are usually lined up with callose, which were shown to be related with the formation of the sieve pores in angiosperms, although not in gymnosperms. Large amounts of callose deposit in the sieve areas also when the sieve element loses conductivity, suffers injury, or becomes dormant. Callose in gymnosperms is typically wound callose. Callose can be easily detected with aniline blue under fluorescence or resorcin blue.

Sieve elements have only primary walls, but sometimes this wall can be very thick receiving the name of nacreous walls and can be present in all major vascular plant lineages. Nacreous walls can be very thick, and they would be secondary walls. Nacreous walls can almost occlude the entire lumen of the sieve element; hence, its presence needs to be considered in experiments of sugar translocation. Such thick walls might be related to resistance to high turgor pressures within the sieve elements. Nacreous walls seem to have a strong phylogenetic signal and are much more common in some families, such as *Annonaceae*, *Calycanthaceae*, and *Magnoliaceae*.

There are basically two types of sieve elements: sieve cells and sieve tube elements. The sieve tube elements are distinguished by the presence of sieve plates, that is, sieve areas with wider and more abundant sieve pores, usually in both extreme ends of the cells, while sieve cells lack sieve plates. A group of connected sieve tube elements form a sieve tube. According to this concept, lycophytes and ferns have sieve cells. However, because of the many differences in the morphology and distribution of protoplasm organelles and chemical substances between the sieve elements of gymnosperms and vascular cryptogams, Evert suggests the use of "sieve cell" as exclusive to the gymnosperms, leaving the more general term "sieve element" to the lycophytes and ferns.

The longevity of sieve elements varies. In many species it is functional for just one growth season, while for other species they can be functional a couple of years, or in the case of plants that lack secondary growth, they will be living for the entire plant life spam. Palm trees would perhaps be the plants with the oldest conducting sieve tube elements, since some reach 200 years. In other plants, on the other hand, the sieve elements collapse a few cells away from the vascular cambium, corresponding to a fraction of the mm. In a mature tree, most of the secondary phloem will generally be composed of sieve elements no longer conducting. This region is called nonconducting phloem, in opposition to the area where sieve elements are turgid and conducting, called conducting phloem. The term collapsed and noncollapsed phloem and functional and nonfunctional phloem are not recommended, since in some plants the nonconducting phloem keeps its sieve elements intact, and although large parts of the phloem may not be conducting, the tissue as a whole is certainly still functioning in storage, protection, and even dividing or giving rise to new meristems, such as the phellogen and the dilatation meristem of some rays.

Sieve Cells and Strasburger Cells

Sieve cells are typically very elongated cells with tapering ends, which lack sieve plates, that is, lack an area in the sieve element where the pores are of a wider diameter. Even though the sieve areas may be more abundant in the terminal parts of the sieve cells, the pores in these terminal areas are of the same diameter as those of the lateral areas of the sieve element. Sieve cells lack P-protein in all stages of development. The sustenance of the sieve cells is carried by specialized parenchyma cells in close contact with the sieve elements, with numerous plasmodesmata, which maintain the physiological functioning of the sieve cells, including the loading and unloading of photosynthates. These cells are known either as albuminous cells or Strasburger cells. The name albuminous was initially coined given the proteinaceous appearance of these cell's contents.

However, because the high protein content is not always present, the name Strasburger cell, paying tribute to its discoverer Erns Strasburger, is recommended over albuminous cells. Strasburger cells in the secondary phloem can be either axial parenchyma cells, as is common in *Ephedra*, or ray parenchyma cells, as is common in the conifers. More commonly, the most conspicuous Strasburger cells in conifers are the marginal ray cells which are elongated and have a larger number of symplastic contact with the sieve cells. Sometimes declining axial parenchyma cells also acts as Strasburger cells in *Pinus*. The only reliable character to distinguish a Strasburger cell from an ordinary cell is the presence of conspicuous connections. In the primary phloem, parenchyma cells next to the sieve cells are those which act as Strasburger cells.

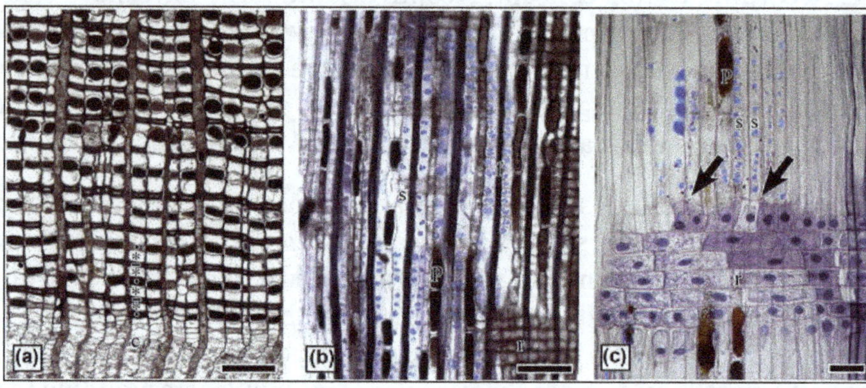

Figure shows The secondary phloem of conifers. (a) Transverse section of the secondary phloem of Sequoia sempervirens (Cupressaceae) showing alternating tangential bands of sieve cells, axial parenchyma, and fibers, interrupted by uniseriate rays. (b). Longitudinal radial section (LR) of the secondary phloem of Sequoia sempervirens (Cupressaceae) showing alternating tangential bands of sieve cells, axial parenchyma, and fibers, interrupted by uniseriate rays. Sieve pores distributed across the walls of long sieve cells. (c). LR section of Pinus strobus (Pinaceae) showing the elongated marginal ray cells in close contact with the sieve cells. These are the Strasburger cells. f and rectangular symbol = fibers, s and * = the sieve cells, p and dot = axial parenchyma cells rich in tannins. Scalebars: a, b = 100 μm, c = 50 μm.

Sieve Tube Elements and Companion Cells

A synapomorphy of the angiosperms is the presence of sieve tube elements and companion cells, both sister cells derived from the asymmetrical division of a single mother cell. In some instances, these mother cells can divide many times, creating assemblages of sieve tube elements and parenchyma cells ontogenetically related. Sieve tube elements have specialized areas in the terminal parts of the sieve elements in which a sieve plate is present. Within the sieve plate, the pores are much wider than those of the lateral sieve areas, evidencing a specialization of these areas for conduction. In *Cucurbita*, the pores in the sieve plate have up to 10 μm in diameter, while the pores in the lateral sieve areas are of about 0.1 μm. The protoplast of sieve tube elements contain a specific constitutive protein called P-protein (P from phloem, also known as slime), which in some taxa (e.g., *Leguminosae*) is nondispersive and can be seen as coagula inside of the sieve element.

Even in lineages of angiosperms where vessels were lost and tracheids re-evolved, such as *Winteraceae* in the *Magnoliids* and *Trochodendraceae* in the *eudicots*, sieve elements and companion cells are present, suggesting the independent evolution of these two plant vascular tissues derived from the same meristem initials. Since the sieve tube element loses its nucleus and ribosomes, the companion cell is the cell responsible for the metabolic life of the sieve elements, including the transport of carbohydrates in and out the sieve elements. Companion cells may be arranged in vertical strands, with two to more cells. Other parenchyma cells around the sieve tube integrate with the companion cells and can also act in this matter. Typically, the cells closely related with the sieve tube elements die at the same time as the sieve element loses conductivity.

Sieve tube elements vary morphologically. The sieve plates can be transverse to slightly inclined or very inclined and contain a single sieve area or many. When one sieve area is present, the sieve plate is named simple sieve plate, while when two to many are present, the sieve plates are called compound sieve plates. Compound sieve plates typically occur in sieve tube elements with inclined to very inclined sieve plates. In addition, sieve elements with compound sieve plates are typically longer than those with simple sieve plates. Evolution to sieve elements of both sieve area types has been recorded in certain lineages, such as in *Arecaceae*, *Bignoniaceae*, and *Leguminosae*, and to the present it is not still clear why the evolution of distinct morphologies would be or not beneficial. The only clear pattern is that compound sieve plates appear in long sieve elements, and phloem with a lot of fibers generally has compound sieve plates.

Parenchyma

In the primary phloem, just one type of parenchyma is present and typically intermingles with the sieve elements. In the secondary structure, there are two types of parenchyma: axial parenchyma and ray parenchyma, derived, respectively, from the fusiform and ray initials of the cambium.

The axial parenchyma in conifers commonly is arranged in concentric, alternating layers. These parenchyma cells contain a lot of phenolic substances, which were viewed as a defense mechanism against bark attackers. In Gnetales, the phloem axial parenchyma appears to be intermingling with the sieve cells. Some of these axial parenchyma cells act as Strasburger cells.

Figure: Phloem axial parenchyma distribution in secondary phloem. (a) Ephedra tweediana (Ephedraceae) TS, showing sieve cells interspersed by axial parenchyma cells (arrows). Six to five cells away from the cambium, the sieve cells already lose conductivity and collapse with axial parenchyma cells enlarging (top arrow). (b) Lannea discolor

(Anacardiaceae) TS showing axial parenchyma with tannins arranged in narrow bands (arrows). There are also other parenchyma cells with less content dispersed in the phloem. Note also the fibers in concentric bands. (c) Robinia pseudoacacia (Leguminosae) TS showing bands of axial parenchyma associated with the fiber bands and sieve tube elements in clusters with simple sieve plates staining with resorcin blue. (d) Fridericia nigrescens (Bignoniaceae) TS with sieve tubes surrounded by sieve-tube-centric axial parenchyma. The tissue background corresponds to the fibers. c, cambium; sx, secondary xylem; c, cambium; sp, secondary phloem; sx, secondary xylem. Scalebars: a = 50 µm; b, d = 200 µm; c = 100 µm.

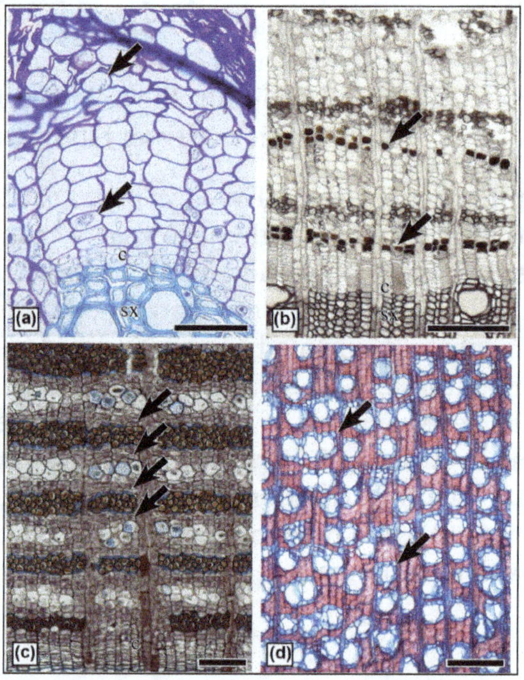

In angiosperms, the distribution of the axial phloem parenchyma is more varied, and it may appear as a background tissue where other cells are dispersed or may be in bands and radial rows or sieve-tube-centric. The distribution of axial phloem parenchyma is commonly related to the abundance of fibers or sclereids. In species with more fibers, it is common to have a more organized arrangement of the parenchyma. For example, in *Robinia pseudoacacia* (*Leguminosae*) there are parenchyma bands in either side of the concentric fiber bands. When very large quantities of sclerenchyma are present, such as in the secondary phloem of *Carya* (*Juglandaceae*) or in *Fridericia*, *Tanaecium*, *Tynanthus*, and *Xylophragma* (*Bignoniaceae*), the sieve-tube-centric parenchyma appears and, as the name suggests, is surrounding the sieve tubes.

Although collectively described and referred to as axial phloem parenchyma, it is important to note that in many plants there will be distinct groups of phloem parenchyma within the phloem with quite different ergastic contents and therefore presumed different functions. Some of these specialized parenchyma cells may be considered secretory structures. Within a single plant, it is not uncommon that while some cells have crystals (especially when in contact with sclerenchyma), others have tannins, starch, and other substances. In apple trees (*Malus domestica, Rosaceae*) three types of axial parenchyma have been recorded: (1) crystal-bearing cells, (2) tannin- and starch-containing cells, and (3) those with no tannin or starch, which integrate with the companion cells.

Within bands of axial parenchyma, canals with a clear epithelium may be formed in many plant groups such as *Pinaceae, Anacardiaceae, Apiales,* a feature with strong phylogenetic signal. Some phloem parenchyma cells also act in the sustenance and support of the sieve elements, even when not derived from the same mother cell. In longitudinal section, the axial phloem parenchyma may appear fusiform (not segmented) or in two up to several cells per strand.

While the phloem ages and moves away from the cambium, its structure dramatically change, and typically axial parenchyma cells enlarge, divide, and store more ergastic contents toward the nonconducting phloem. In plants with low fiber content, the dilatation undergone by the parenchyma cells typically provokes the collapse of the sieve elements. The axial parenchyma in the nonconducting phloem can dedifferentiate and give rise to new lateral meristems. In plants with multiple periderms, typically new phellogens are formed within the secondary phloem, compacting within the multiple periderms large quantities of dead, suberized phloem. In plants with variant secondary growth, especially lianas, new cambia might differentiate from axial phloem parenchyma cells. In the Asian *Tetrastigma* (*Vitaceae*), new cambia were recorded differentiating from primary phloem parenchyma cells.

Sclerenchyma

Sclerenchymatic cells are those with thick secondary walls, commonly lignified. Sclerenchyma can be present or not in the phloem, and when present it typically gives structure to the tissue. For instance, a phloem with concentric layers of sclerenchyma cells is called stratified, regarding the organization of the elements in tangential section. In Leguminosae, bands of phloem are associated to the concentric fiber bands.

Older phloem shows more sclerification than younger phloem, and the sclerenchyma may also act as a barrier to bark attackers. The sclerenchyma is typically divided in two categories: fibers and sclereids. These cell types differ mainly in form and size, but origin has also been used to distinguish them.

Fibers

Fibers are long and slender cells, derived from meristems, the fiber primordia. In the primary phloem, fiber caps are sometimes found in association with the protophloem and are named protophloem fibers. Since only an ontogenetic study can evidence whether these fibers indeed differentiate within the protophloem, a term coined, pericyclic fibers, has been recommended to be used instead of primary phloem fibers or perivascular fibers. In the monocotyledons, fibers are commonly an important component of the vascular bundles. Commonly these fibers are associated with the phloem, but they might also be associated with the xylem or be central in the vascular bundle. These fibers are not, however, understood as part of either phloem or xylem; although they are of vascular nature, they differentiate directly from procambium.

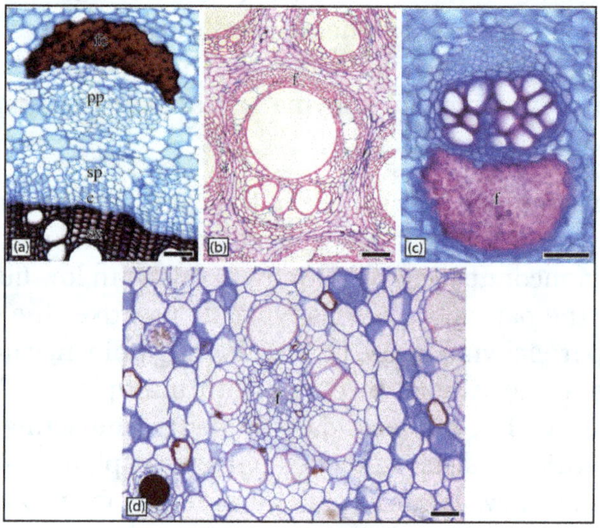

Figure shows Vascular fibers associated to eudicot and monocot primary structure. (a) Peri-cyclic fiber cap (fc) and primary phloem (pp) in Perianthomega vellozoi (Bignoniaceae). Secondary phloem (sp) beginning to be produced. Vascular bundles in monocot-yledons. (b) Vascular bundle in the climber Calamus manan (Arecaceae) with fibers toward the phloem side. Phloem in two strands around a wide metaxylem vessel. (c) Vascular bundle of Vellozia alata (Velloziaceae), with fiber cap toward the xylem side. Phloem on the top side of the picture. (d) Amphivasal bundle of Philodendron with fibers in the center of the vascular bundle and phloem surrounding it. Scalebars: a, b = 100 µm, c, d = 50 µm.

Sclereids

Sclereids may have different forms and sizes. Within the phloem, they are more typically square or polygonal (stone cells) and contain numerous pits and conspicuous pit canals. Holdheid defines that a sclereid is a cell derived from the belated sclerification of a parenchyma cell, and that is in fact the rule in the majority of cases. However, there are lineages in which the sclereids differentiate very close to the cambium and it would be untrue to claim that the derivatives had a stage as a mature parenchyma cell. In these cases, the form is enough to define the sclereid.

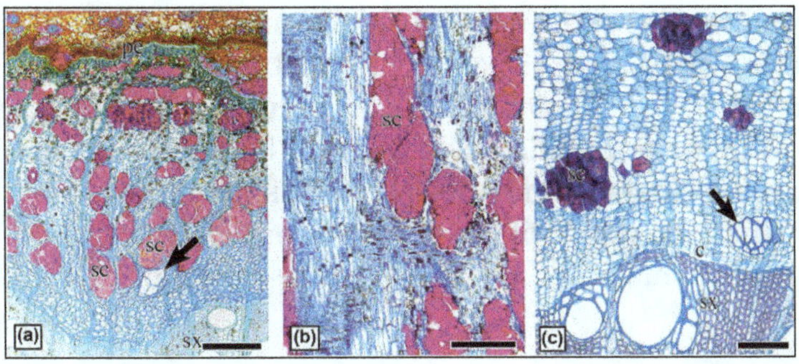

Figure shows Sclereids in the secondary phloem. (a) Sclereids (sc) differentiate from parenchyma cells (arrow) in the nonconducting phloem of Heteropterys intermedia (Malpighiaceae) TS, forming large clusters. (b) Longitudinal radial section of Heteropterys intermedia (Malpighiaceae) showing the sclereid masses. (c) In Pleonotoma tetraquetra (Bignoniaceae), the sclereids differentiate (arrow) close to the cambium within the conducting phloem. c, cambium; pe, periderm; sc, sclereid; sx, secondary xylem. Scalebars: a, b = 400 µm, c = 250 µm.

On the other hand, there are cases where long and slender cells derive from previously mature parenchyma cells and are morphologically difficult to distinguish from fibers. In these cases, these cells are called fiber sclereids and may be even in concentric layers, such as in apple trees and pears (*Malus domestica* and *Pyrus communis*, respectively). Sclereids can also develop with different arrangements in the phloem, being isolated and scattered or in clusters c.

Rays

The rays in the conducting phloem have typically the same organization in terms of width, height, and cellular composition as the secondary xylem. In this respect the rays vary from uniseriate to multiseriate and may be homocellular or heterocellular. Homocellular rays are those composed of cells of one shape, all procumbent or all upright (common in many shrubs). Heterocellular rays are those where more than one cell shape is present together. Ray composition is appreciated in radial sections.

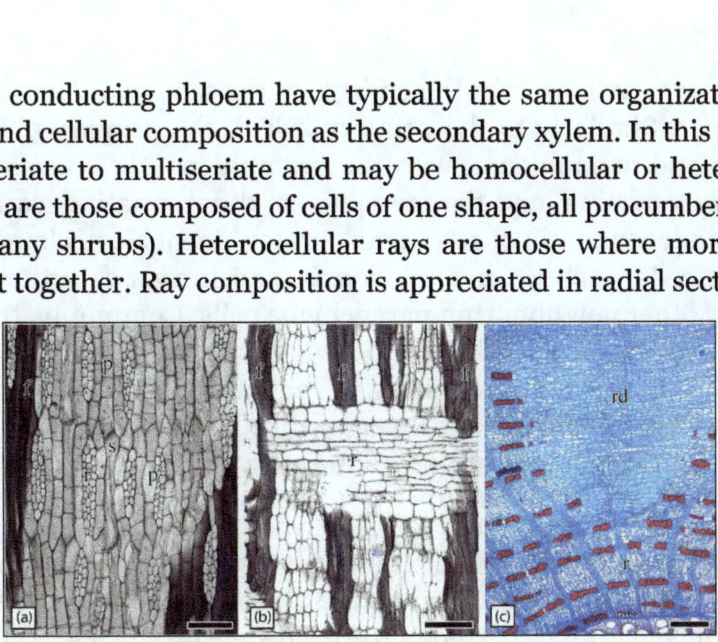

Figure: Rays in the secondary phloem. (a) Longitudinal tangential section of Brachylaena transvaalensis (Asteraceae) showing storied structure, biseriate to triseriate rays (r), sieve tube elements with simple sieve plate (s) and axial parenchyma cells composed of 4–5 cells, and fibers. (b) Longitudinal radial section of Brachylaena transvaalensis (Asteraceae) showing heterocellular rays (r), with body procumbent and one row of marginal square cells. Fibers (f) in bands (c) Ray dilatation (rd) by the formation of a dilatation meristem in the center of the ray in Perianthomega vellozoi (Bignoniaceae). f, fiber; p, axial parenchyma cell; r, ray; s, sieve tube element. Scalebars: a, b = 100 µm, c = 300 µm.

Because the vascular cambium produces much more xylem to the inside than phloem to the outside, phloem rays typically greatly dilate toward the periphery of the organ. It is not uncommon that a dilatation meristem longitudinal to the cambium forms in some barks, especially in families with very wide, wedge-like rays such as the *Malvaceae*. Plants with unicellular rays very rarely have dilatation by cell division. Instead, they have great lateral expansion of their single cells. Ray width can be only determined in tangential sections.

Rays are typically exclusively parenchymatic; however, in many species sieve elements appear in the rays and are called ray sieve cells or radial sieve cells. These cells were recorded connecting two different sieve tubes (collections of sieve tube elements). Ray sieve elements seem to be present in taxa where perforated ray cells have been also recorded.

Structure and Development of Primary and Secondary Phloem

Primary Phloem

The primary phloem derives from the embryo in the seed and the procambium from the organ's apices. Similarly to the primary xylem, the primary phloem is divided in protophloem and metaphloem, with the protophloem differentiating first, while the plant is still elongating, and the metaphloem differentiating last. The phloem is always exarch, independently of the organ. Protophloem sieve elements sometimes lack companion cells, such as in *Arabidopsis*, and in this case the sieve elements are sustained by other neighbouring parenchyma cells. Commonly, the protophloem quickly becomes obliterated and loses function. In plants without secondary growth, the metaphloem will be conducted during the entire life of the plant, as in the monocotyledons. Different vascular plant lineages display different arrangements of the primary xylem and phloem, depending on the stele type.

Two main types of steles exist, the protostele and the siphonostele. In the protostele, the entire center of the organ is composed of vascular tissue, with the phloem in strands alternated with a central xylem in the protostele, haplostele, and actinostele, while primary phloem is interspersed in the protostele plectostele. The roots of all the vascular plants are protostelic. The stems, however, can vary. In the lycophytes, they are always protostelic, while in the ferns (monilophytes) they might be protostelic, such as in *Psilotum*, or in all other range of siphonostelic steles. The siphonostele evolved in concert with the macrophytes and resulted in the formation of a central pith derived from the ground meristem. No lineage displays as much diversity in the primary vasculature architecture as do the ferns. In the seed plants, that is, gymnosperms and angiosperms, the stem stele is always a syphonostele, either a eustele, where discrete vascular bundles form a concentric ring, or the atactostele, a type of stele exclusive of the monocotyledons where the bundles are scattered in the entire stem center. Some lineages of eudicotyledons and *Magnoliids* have evolved another subtype of siphonostele, the polycyclic eustele, where more than one ring of bundles is present, such as in *Piperaceae* and *Nyctaginaceae*.

The primary phloem is simpler than the secondary phloem and is basically formed by sieve elements and parenchyma cells. Fiber caps are commonly present, and they might be phloematic. The position of the phloem is typically external or abaxial to the xylem, but in some lineages the bundles are bicollateral, and phloem is present both inside and outside (abaxial and adaxial), while in amphivasal bundles, the xylem encircles the phloem, as in the secondary vascular tissues of some *Asparagales* and *Iridaceae* corms. In some plant families and orders, intraxylary phloem (perimedullar phloem islands) is a synapomorphy, such as in the order *Myrtales* and in the families' *Apocynaceae* and *Convolvulaceae*. These phloem strands are initially primary, but a cambium can differentiate between the protoxylem and the phloem strands and develop secondary tissues inside of the pith.

Secondary Phloem

Being derived from the cambium, the secondary phloem will share a number of characteristics with the secondary xylem. For instance, it is divided in an axial and radial system. The axial system is composed of sieve elements, axial parenchyma cells, and fibers, and the radial system is formed by rays, which are typically parenchymatic. Similar to secondary xylem, the secondary phloem can be storied or non-storied, depending whether the cambial mother cells are organized in tiers or not.

Some trees will have growth rings, with an early and a late phloem, both in temperate and tropical regions, but their characterization is only possible with periodical collections. Sometimes, but not always, the fiber band width gives a hint on the presence of growth rings or the formation of very small sieve elements in the late phloem.

Secondary Phloem of Gymnosperms

In conifers (except Gnetales) the secondary phloem is typically marked by an alternation of axial cell types, uniseriate rays, and, in many lineages, axial and radial resin canals (e.g., *Pinaceae* and *Cupressaceae*). In the *Pinaceae*, the phloem is marked by the presence of an alternation of sieve cells and bands of axial parenchyma with phenolic contents, some also with druses. In the nonconducting phloem of *Pinaceae*, sclereids differentiate. In all other conifers, in addition to the alternation of parenchyma bands and sieve cells, fiber bands are present. Therefore, sieve cells, parenchyma cells with phenolic content, and bands of fibers appear in alternation in non-*Pinaceae* and Gnetales conifers, including *Araucariaceae*, *Cupressaceae*, *Podocarpaceae*, *Taxaceae*, and *Taxodiaceae*. Another marked difference of these conifers compared to *Pinaceae* is that they contain a lot of crystals in their cell walls, including in Gnetales, while in *Pinaceae* they are exclusively inside of idioblastic cells.

In other gymnosperms, in particular in Gnetales and Cycads, the first remarkable difference is the presence of very wide, multiseriate rays alternating with uniseriate rays. The wide rays in both groups have, however, evolved independently, since Cycads are a sister to all other gymnosperms, while Gnetales are within the conifers, as sister to the Pinaceae. In *Cyca* and the extinct *Cycadoidea*, sieve cells and phloem parenchyma alternate with fibers, which can be in tangential bands or not. In *Cyca*, the sieve cells appear in radial rolls, while in *Cycadoidea* there is a constant alternation of one sieve cell or phloem parenchyma to one fiber. The nonconducting phloem of *Cycas* is marked by the collapse of sieve cells, enlargement of the axial parenchyma cells, ray dilatation, and sclerosis of some parenchymatic cells. More than one ring of secondary phloem is present in some Cycads (e.g., *Cycas*, *Encephalartos*, *Lepidozamia*, and *Macrozamia*) and Gnetales (e.g., *Gnetum*), given that they have successive cambia.

Within the Gnetales, in *Ephedra* axial parenchyma cells are interspersed with sieve cells, and fiber may or may not be present and are typically gelatinous. Fiber sclereids and/or sclereids appear in the nonconducting phloem of other species. In the nonconducting phloem of *Ephedra*, the sieve cells and Strasburger cells collapse with the enlargement of the axial and radial parenchyma cells with more ergastic contents. In *Gnetum*, large areas of parenchyma sclerify, forming bands in the nonconducting phloem. The secondary phloem of *Welwitschia* is described as containing a large amount of fibers.

Secondary Phloem of Angiosperms

Within the angiosperms, the diversity of phloem cell type arrangements reaches its maximum. The structure can be storied or non-storied; sclerenchyma can be present or lacking. The rays may be uni-, bi-, or multiseriate. A large array of secretory cells may be encountered, such as resin canals, laticifers, and mucilaginous cells. Crystalliferous parenchyma is also very common, especially when associated with fibers.

The variation in cell type arrangements can be of taxonomic interest. Sieve elements can vary in morphology and arrangement. They can be solitary, scattered in the phloem (e.g., *Eucalyptus*, *Myrtaceae*), in clusters, and in radial or tangential rows. The functional significance of the different arrangements is unknown to date, although this is one of the features in the phloem with the strongest phylogenetic signal.

The presence, type, and arrangements of fibers and sclereids are one of the most informative characters in the bark. In *Apocynaceae*, the fibers are completely absent, except in *Aspidosperma*, the sister group of all other *Apocynaceae*. In *Aspidosperma*, they can appear solitary scattered across the phloem or in clusters. In some lineages, fibers appear in concentric alternating bands, as in *Leguminosae* (*Papilionoideae*), *Mimosoideae*, *Bignoniaceae*, and *Malvaceae*, and this is a constant character among them. Phloem parenchyma more commonly constitute the background tissue in the phloem but can also be distributed in bands, radial rows, or even only around the sieve tube elements.

Phloem Activity

The classic theory of phloem transport is that proposed by Ernst Münch, and it involves the formation of an osmotic pressure transport gradient, where certain zones act as sources of sugars (leaves and storage organs), while others act as sinks. Experiments showed that the concentration gradients were always seen to be positive in the direction of flow, supporting Münch's postulate. In a system where transport goes against the direction of transpiration, its functionality relies on the presence of a plasma membrane across the entire system to create an osmotic pressure, hence the need of a conducting system with living cells. Recent studies have been refining aspects involved in the photosynthate conduction to explain long-distance transports across large trees with such a simple system. A direct role of intracellular calcium has also been reported in the dissolution of nondispersive P-proteins and facilitation of transport. For instance, phloem sieve element length scale with the tree sizes and sieve plate type. It was also shown that sieve element's diameter, length, and pore width increase from the top to the base of the trees.

Across the entire pathway, sugars are removed from the system to sustain all cells in the plant body. This mechanism is only possible with the concerted mechanism between sieve elements and their close related cells (Strasburger cells and companion cells), with these accompanying cells constantly channeling substances and macromolecules toward the sieve elements. The Strasburger and companion cells carry the loading and unloading of the sieve elements. Given the function of loading and unloading, the companion cell-sieve tube element size ratio is directly related to being in the source or the sink of sugars. For instance, in leaves the companion cells are typically much larger, for they have the high demand of constantly loading the sieve tubes. In areas of release of the sugars (unloading), the companion cells are much smaller or even absent.

Economic Uses

In the economic uses, it is not always easy to distinguish the use of the phloem from that of the periderm, since both together compose the bark of a woody plant. The phloem corresponds to the inner bark, and the periderm to the outer bark. The bark has a long history of utilization, from the production of remedies, aphrodisiacs (yohimbe), insecticides, dyes, tannins, angostura, fibers, gums and resins, latex, and flavorings.

In indigenous groups from British Columbia (Canada) and Tanzania, barks from dozens of species of woody plants are used as carbohydrate food, medicine, fibers, and structural material. In Mexico the bark of *Ficus* is used since prehispanic times to create a type of paper called *papel amate* (from the náhuatl paper = ámatl), used, for example, to create the Aztec codices.

The rubber tree, *Hevea brasiliensis* (*Euphorbiaceae*), is known from the extraction of latex to the production of rubber. Laticifers are present in concentric rings in the secondary phloem of the rubber tree and are an important economic asset in some tropical countries. Bark residues have also been considered for mulching, to build particle boards, as fuel, and a source of food for ruminants.

References

- Meristem-cells: microscopemaster.com, Retrieved 13, August 2020
- Parenchyma-cells: biologydictionary.net, Retrieved 16, March 2020
- Collenchyma, science: britannica.com, Retrieved 20, July 2020
- Collenchyma-tissue-in-plants-simple-tissue, simple-tissue, plant-tissues-69066: biologydiscussion.com, Retrieved 28, January 2020
- Epidermal-cells: microscopemaster.com, Retrieved 02, June 2020
- Phloem-cell-types-structure-and-commercial-uses, plant-science-structure-anatomy-and-physiology-in-plants-cultured-in-vivo-and-in-vitro: intechopen.com, Retrieved 25, February 2020

Chapter 3

Nucleus

Nucleus is the membrane-bound organelle which encloses the genetic material of eukaryotic organisms. Eukaryotic cells generally have a single nucleus but some cells such as red blood cells lack nucleus and some cells, for example, osteoclasts have multiple nuclei. This chapter has been carefully written to provide an easy understanding of the varied facets of nucleus such as nuclear envelope and nuclear pore complex.

Nucleus is the most important and highly specialized organelle of the cell which contains genetic information necessary for cellular metabolism, growth and differentiation. The 'nucleus' was first described in orchids in 1831 by a Scottish Botanist, Robert Brown as "single circular structure which was more opaque than the cell membrane." Nucleus is present only in eukaryotic cells. Generally there is only one nucleus per cell, but some exceptions exist like siphonales group of algae, where there are more nuclei present. Typical nucleus occupies about 10% of a eukaryotic cell's volume. The nucleus mainly contains all the genes, collectively called as nuclear genome. The size of the nuclear genome in plants is highly variable ranging from 1.2×10^5 Kb in Arabidopsis thaliana to 1×10^8 Kb in a lily, Fritillaries. Some of the genetic information is also contained in two organelles viz. chloroplast and mitochondria.

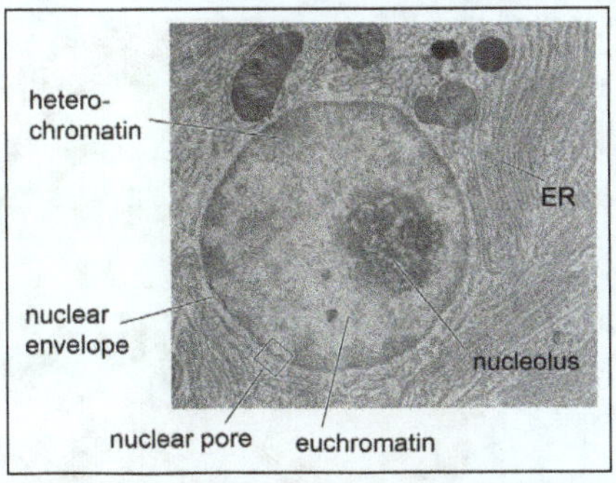

Transmission electron micrograph of a plant cell showing nucleolus, chromatin and nuclear envelope.

Constituents of Nucleus

Nuclear Envelope

The nuclear membrane, also called the nuclear envelope, is a double membrane layer that separates the contents of the nucleus from the rest of the cell. It is found in both animal and plant cells. A cell has many jobs, such as building proteins, converting molecules into energy, and removing waste products. The nuclear envelope protects the cell's genetic material from the chemical reactions that take place outside the nucleus. It also contains many proteins that are used in organizing DNA and regulating genes.

Function of the Nuclear Membrane

The nuclear membrane is a barrier that physically protects the cell's DNA from the chemical reactions that are occurring elsewhere in the cell. If molecules that stay in the cytoplasm were to enter the nucleus, they could destroy part of the cell's DNA, which would stop it from functioning properly and could even lead to cell death. The envelope also contains a network of proteins that keep the genetic material in place inside the nucleus.

It also manages what materials can enter and exit the nucleus. It does so by being selectively permeable. Only certain proteins can physically pass through the double layer. This protects genetic information from mixing with other parts of the cell, and allows different cellular activities to occur inside the nucleus and outside the nucleus in the cytoplasm, where all other cellular structures are located.

Parts of the Nuclear Membrane

The nuclear membrane surrounds the nucleus of the cell.

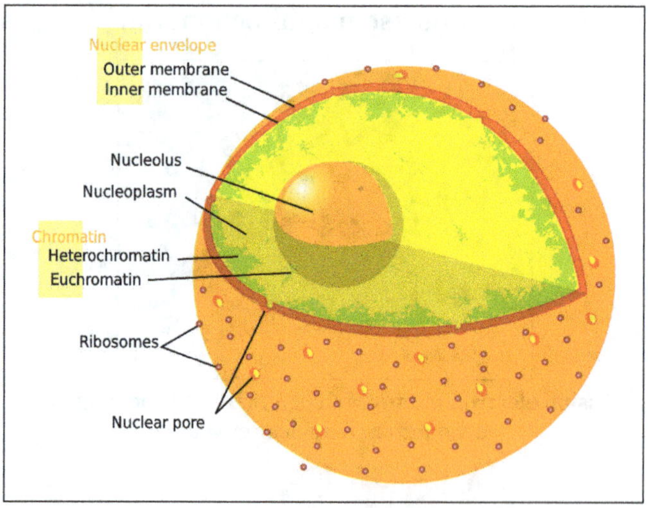

Outer Membrane

Like the cell membrane, the nuclear membrane is a lipid bilayer, meaning that it consists of two layers of lipid molecules. The outer layer of lipids has ribosomes, structures that make proteins, on its surface. It is connected to the endoplasmic reticulum, a cell structure that packages and transports proteins.

Inner Membrane

The inner membrane contains proteins that help organize the nucleus and tether genetic material in place. This network of fibers and proteins attached to the inner membrane is called the nuclear lamina. It structurally supports the nucleus, plays a role in repairing DNA, and regulates events in the cell cycle such as cell division and the replication of DNA. The nuclear lamina is only found in animal cells, although plant cells may have some similar proteins on the inner membrane.

Nuclear Pores

Nuclear pores pass through both the outer and inner membranes of the nuclear membrane. They are made up of large complexes of proteins and allow certain molecules to pass through the nuclear membrane. Each nuclear pore is made up of about 30 different proteins that work together to transport materials. They also connect the outer and inner membranes.

During cell division, more nuclear pores are formed in the nuclear membrane in preparation for cell division. The nuclear membrane eventually breaks down and is reformed around the nuclei of each of the two daughter cells. The figure below shows a nuclear pore close-up:

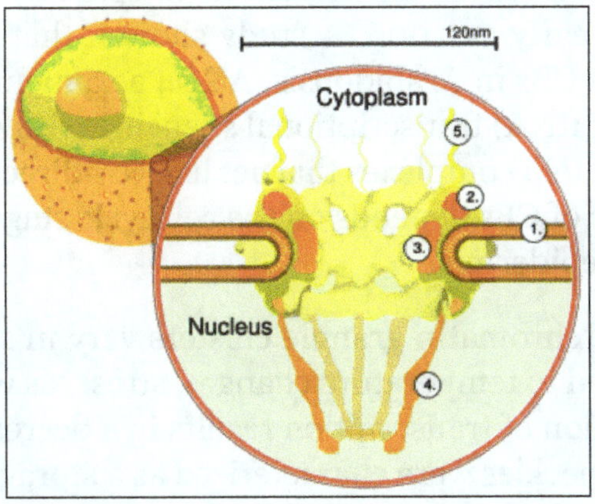

Nuclear Bodies

Nuclear bodies are common and dynamic structures of eukaryotic nuclei. A majority are localized outside the nucleolus, some inside it, and sometimes both options are possible. However, information on the mechanisms by which nuclear bodies are formed and maintained, and how cells regulate their assembly process, has not yet been provided.

Plant nuclear bodies are membrane-less subnuclear organelles comprising the nucleolus, CBs, nuclear speckles, cyclophilin-containing speckles, dicing bodies, AKIP1-containing bodies, and photobodies.

Nucleoli, the largest bodies in eukaryotic interphase cell nuclei, are known as producers of ribosomal subunits and participate in many fundamental cellular activities. The plant nucleolus shows a well-ordered structure with various components such as fibrillar centres (both heterogenous and homogenous), a dense fibrillar component, a granular component, nucleolar vacuoles, nucleolar chromatin, and nucleolonema. Nucleoli are mostly composed of proteins (85–90%). RNA represents only 5–10% and the smallest part is rDNA. When cells enter mitosis, some nucleolar proteins diffuse into the cytoplasm and some remain attached to the nucleolus organizing region (NOR) of mitotic chromosomes. A few nucleolar proteins have already been characterized in plants, including several nucleolin homologues or nucleolin-like proteins, and many ribosomal proteins. In addition, using a proteomic approach, 217 nucleolar proteins of Arabidopsis have been identified.

CBs are very dynamic particles. They can move within the nucleus, into the nucleolus and fuse together. A major component of CBs is the protein coilin, which serves as a marker for this structure. The plant homologue of coilin, Atcoilin, shows a strong affinity for snRNA (small nuclear RNA) as well as for non-specific RNA. In addition, a characteristic feature of plant cells is the presence of poly(A) mRNA including protein-coding RNAs in CBs. Thus, the function of plant CBs could be connected with the storage or retention of poly(A) RNAs. The survival of motor neurons protein (SMN) is another CB protein involved in the biogenesis of splicing small nuclear ribonucleoproteins (snRNPs). As early as 2006, A study showed that ARGONAUTE4 (AGO4) protein co-localized with CBs in Arabidopsis. AGO4 as a member of the 24 nucleotide siRNA (small interfering RNA) transcriptional silencing pathway could function in the generation of siRNA–protein complexes that act in RNA-directed DNA methylation. In Arabidopsis, the number of CBs depends on the cell cycle stage, cell type, and developmental stage of the root epidermis.

Nuclear speckles or interchromatin granule clusters vary in size, shape, and number. These dynamics are linked to temperature changes, stress, as well as to transcription or phosphorylation. Inhibition of transcription results in a decrease in the number of nuclear speckles. Nuclear speckles were characterized as a storage site for splicing factors and were detected near active transcription sites. Even though the protein composition of metazoan nuclear speckles has already been analysed, the components of plant nuclear speckles remain elusive.

Plant-specific Nuclear Bodies

A study showed that arginine/serine-rich (RS) domain-containing cyclophilin (CypRS64), which interacts with Ser/Arg-rich proteins as important splicing factors, localized to a small number of nuclear bodies different from CBs—cyclophilin-containing speckles.

Dicing bodies are comprised of microRNA processing proteins DCL1 (dicer-like1) and

HYL1 (hyponastic leaves1) and are required for processing of pri-miRNA (primary microRNA) and/or storage/assembly of the miRNA processing machinery.

AKIP1-containing bodies were observed in 2002 by Li et al. AKIP1–green fluorescent protein (GFP) was localized in the guard cell nuclei and was relocalized to the speckles after treatment with abscisic acid. Photobodies are unique to plants and are regulated by an external light signal.

Nuclear Pore Complex

Nuclear pore complexes are very abundant in cells showing an intense trafficking between nucleoplasm and cytoplasm, as in differentiating cells. It is estimated around 11 nuclear pore complexes per μm2 of nuclear envelope, which means around 3000 to 4000 nuclear pores in a single cell. In the cell cycle, new nuclear pore complexes are synthesized and assembled during interphase, previously to mitosis. However, they are also synthesized after cell division. In open mitosis, where the nuclear envelope is disorganized, the nuclear pore complexes are also disorganized and the their proteins are freed in the cytoplasm. After mitosis, these nuclear pore proteins are grouped again in new nuclear pore complexes while the nuclear envelope is assembled at the same time.

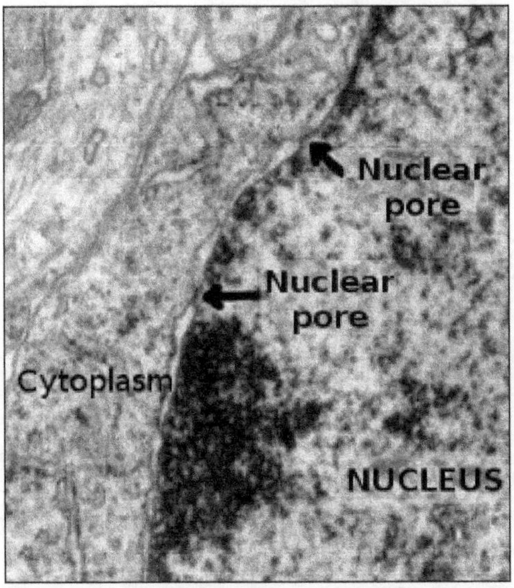

Transmission electron microscopy image of the nuclear envelope. The two constrictions of the nuclear envelope are nuclear pore complexes.

Components

Proteins of the nuclear pore complexes are known as nucleoporins. In yeasts, 30 different nucleoporins have been found, whereas in metazoa may be more than 40 different nucleoporins. There are several copies of each nucleoporin in the same nuclear pore complex. Thus, in mammals, a nuclear pore complex contains around 500 to 1000 nucleoporins. The total structure is 100 to 150 nm wide, 50 to 70 nm high, and contains an inner hydrophilic passage of around 40 nm. Nuclear pore complexes are among the largest protein structures of the cell, about 125000 kDa. It is remarkable that nucleoporins show a great stability, sometimes as long as the whole life of the cell, whereas the lifetime of a common protein may last a few days.

Nucleoporins are grouped in 8 blocks, which are organized as a regular octagon forming ring-like structures. The cytoplasmic ring faces the cytoplasm, the radial ring is in the channel of the nuclear envelope and anchors the nuclear pore complex to the nuclear envelope, and the nuclear ring is facing the nucleoplasm. Furthermore, there are fibrils extending from each of the 8 blocks: cytoplasmic fibrils and intranuclear fibrils. The intranuclear fibrils are connected to the nuclear ring by one of their ends and to other proteins that form another intranuclear ring known as distal ring. Intranuclear fibrils and distal ring together form the nuclear basket, also known as nuclear cage.

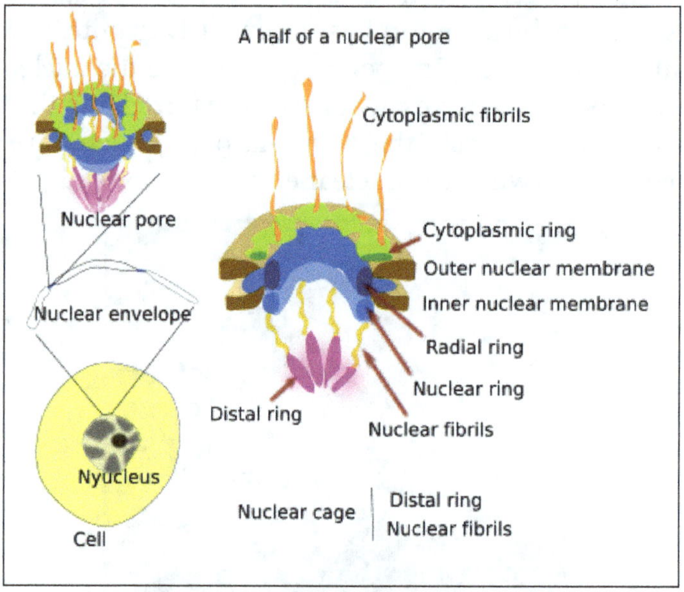

Protein organization in a nuclear pore complex.

Besides the structural role, nucleoporins are classified according to their function. There are trans membrane proteins for anchoring the whole complex to the membrane of the nuclear envelope. The scaffold proteins form the rings, and inner proteins that form the hydrophilic passage and regulate the molecular trafficking through the nuclear pore complex. Those proteins that form the fibrils and nuclear cage recognize the molecules that are then allowed to cross the nuclear passage. It should be noticed that molecules coming in and going out of the nucleus do no need to cross any membrane, but just the hydrophilic channel of the nuclear pores.

Nucleo-cytoplasm Transport

The nuclear pore complex contains a hydrophilic passage ranging from 80 to 90 nm in diameter. When a nuclear pore complex is at rest (no trafficking) the usable space is about 45 to 50 nm in diameter, and it can be increased when some molecule is being transported. Small molecules (less than 60 kDa) like salts, nucleotides, small molecules and short polypeptides can cross freely through the hydrophilic channel, but other larger molecules with physiological roles are not allowed to go across freely. Even some molecules smaller than 20-30 kDa such as histones, tRNAs and small mRNAs may need the involvement of nucleoporins to cross the nuclear envelope. The selective transport, mediated by nucleoporins, is known as passive facilitated transport. Energy is not needed for it. The nuclear pore complex does not determine the direction of the transport, getting in or out the nucleus, but molecules travel according to a gradient of the Ran proteins. The energy is spent in generating this Ran gradient.

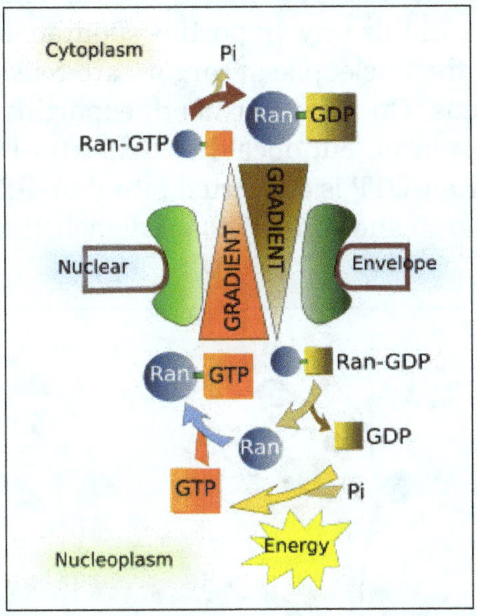

Ran gradient between the cytoplasm and nucleoplasm. In the cytoplasm, the energy needed to create this gradient is supplied by ATP, transforming Ran-GDP in Ran-GTP. Thus, the nucleoplasm is a sink of Ran-GDP and a source of Ran-GTP. In the cytoplasm, Ran-GTP is converted into Ran-GDP. Thus, the cytoplasm is a source of Ran-GDP and a sink of Ran-GTP. In this way two gradients are created, Ran-GDP and Ran-GTP. The size of the icons in the figure depicts levels of concentration.

Trafficking through a nuclear pore complex is high, with more than 1000 translocations per second. This movement of molecules across the nuclear envelope is regulated by the gradient of Ran proteins. Ran are involved in both importing and exporting molecules between the nucleus and the cytoplasm. They generate the Ran-GTP/Ran-GDP needed for the transport, and the generation of these gradients consumes ATP. Ran molecules can be in three states: Ran-GTP, Ran-GDP and Ran. The state of a Ran molecule depends on several enzymes. In the nucleoplasm, Ran-GDP is more abundant, whereas Ran-GTP is concentrated in the cytoplasm.

Karyopherins are a family of proteins, divided in two subfamilies: importins and exportins, which are responsible for selecting the molecules that can cross the nuclear pore complex. Proteins that need to be imported into the nucleus have a particular amino acid sequence, known as entrance signal peptide, and those that need to be exported to the cytoplasm have an exit signal peptide. These short peptide sequences are not identical for all proteins, but similar. The entrance and exit signal peptides are recognized by importins and exportins, respectively. There are members of importins and exportins with different affinity for the import and exit sequences. Nucleoporins do not interact directly with the transported molecules, but with importins and exportins.

Importin and exportin use the Ran-GTP/Ran-GDP gradients for transporting the cargoes in a specific direction. In this way, importins spontaneously joins to their cargoes in the cytosol, but once in the nucleoplasm cargoes are released by Ran-GTP, which is abundant inside the nucleus. On the other hand, exportins need Ran-GTP to join to their cargoes in the nucleoplasm, but once they (exportin-cargo-Ran-GTP) are translocated to the cytoplasm, Ran-GTP is converted into Ran-GDP, which breaks the complex, and then exportin, cargo and Ran-GDP are detached from each other becoming free in the cytosol.

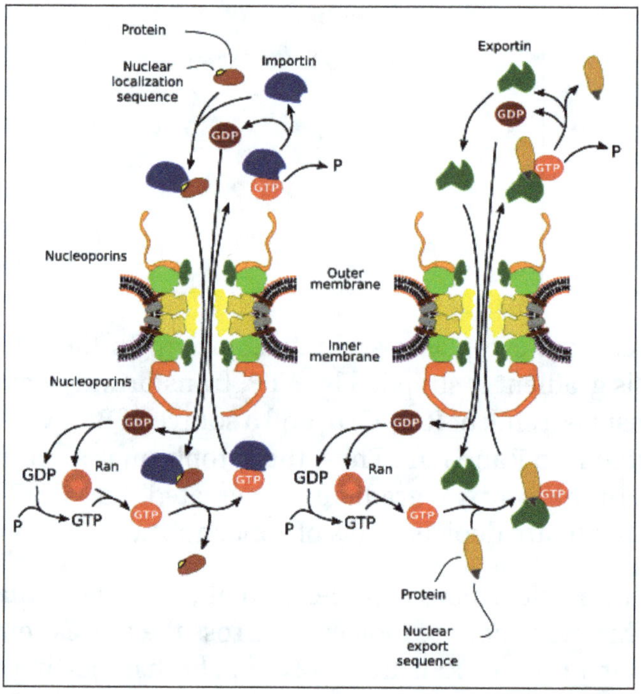

Molecular trafficking across the nuclear pore complex mediated by karyopherins: Importins and Exportins.

Molecules to be specifically transported across the nuclear envelope need to have export or import sequences, but it is not enough. These sequences need to be accessible to exportins and importins. Conformational changes or chemical modifications of the molecules having these sequences may prevent their recognition by karyopherins, so they remain in the nucleoplasm or cytoplasm until the sequence is properly exposed. This mechanism adds a new step of regulation to nucleus-cytoplasm trafficking.

Besides proteins, other molecules need to cross the nuclear envelope. Different types of RNA are transported by different mechanisms, but there are always proteins involved. mRNA is not transported naked. Once the mRNA transcript is synthesized, it is quickly associated with different nuclear proteins forming a large size molecular complex known as ribonucleoprotein. Before mRNA is transported, a quality control checks if it has been correctly processed. If there are any errors, the mRNA is degraded and it does not reach the cytoplasm. If there are no flaws, the ribonucleoprotein is transferred to the nuclear basket of the nuclear pore complex and gets associated with the proteins Nxf1-Nxt1, which facilitate its transport through the hydrophilic channel. This transport does not use the Ran-GTP gradient. However, it needs the hydrolysis of ATP, which is consumed when the macromolecular association (ribonucleoprotein-Nxf1-Nxt1) needs to be disengaged in the cytosolic face of the nuclear pore complex. Then, ribonucleoprotein is free from Nxf1-Nxt1 and cannot get back to the nucleus. Some mRNA molecules uses the protein CRM1 to be transported. In this case, the Ran-GTP gradient is needed.

Transfer RNA (tRNA) is recognized by a type of exportin known as exportin-t, which uses the Ran-GTP gradient to get the tRNA out from the nucleus. Small nuclear RNA (snRNA), another type or RNA, is transported by CRM1 protein and Ran-GTP gradient. The exporting mechanism for ribosomal subunits, which are assembled in the nucleolus, is a big challenge for nuclear pore complexes because they are really large protein-rRNA complexes. The molecular changes happening in the nuclear pore complex allowing the ribosomal subunits to get to the cytoplasm is not known so far. Finally, it has been shown that some proteins are able to bind directly to nucleoporins, so they do not need karyopherins.

Chromatin Interactions

Nuclear pore complexes are involved in additional functions that help in the regulation of the nucleus-cytoplasm communication. In this context, they are thought to participate in DNA repairing and in the regulation of transcription, processing and quality control of mRNA. At transmission electron microscopy, it is observed that heterochromatin is not distributed near the nuclear pore complexes. Thus, they are regarded as spots of permissive gene expression of many genes. This would be an advantageous position for those genes since their transcribed mRNAs are released close to the gateway. The lack of heterochromatin near the nuclear pore complexes may be a consequence of direct interactions between nucleoporins and chromatin.

Nucleoporins are evolutively related to COPI and COPII proteins, which form the coated vesicles coming from the Golgi apparatus and endoplasmic reticulum, respectively. It makes sense because the nuclear envelope and endoplasmic reticulum membranes are continuous, and Golgi apparatus membranes are formed from vesicles coming from the endoplasmic reticulum.

Nuclear Import and Export

The main function of nuclear pores is to carry out diffusion of small molecules and active transport of macromolecules across nuclear membrane. Molecules smaller than 30 KDa can readily diffuse through the nuclear pores. Molecules of size up to 60 KDa can be transported through the pore passively. Larger molecules require active transport. Several million macromolecules selectively pass through the nuclear pore every minute. Active transport is carried out by the Ran cycle and transport receptors, for those molecules, which have nuclear localization signal.

Nuclear Import

Selective transport of proteins into the nucleus is an essential process in eukaryotic cells. The transport involves recognition of specific cargoes in the cytoplasm by soluble import receptors and subsequent transport of the cargo/receptor complex into the nucleus through nuclear pore complexes (Wu et al, 2009). Many different protein are imported into the nucleus from cytoplasm e.g. histones, DNA polymerases, RNA polymerases, transcription factors, splicing factors etc. These proteins are targeted to nucleus by a specific amino acid sequence called as nuclear localization signal (NLS). These NLS are recognized by the transport receptors and the protein bearing these signals is transported through nuclear pore complex (NPC). Three major types of NLS are identified in plants. The components controlling active transport are highly conserved in animal, plant and fungal kingdoms and they all show lot of homology.

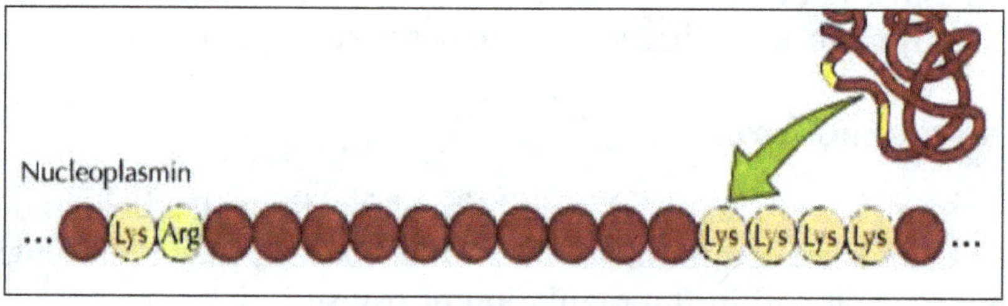

Bipartite NLS of nucleoplasmin.

Nuclear import in plants shows unique features than other eukaryotes. In animals, importin α serves as receptor for NLS on the cargo protein and importin β interacts with the NPC. So formation of a complex of cargo protein + importin α + importin β, in the cytoplasm is very much necessary for transport across the membrane in the nucleus. So in damaged animal cells, nuclear import is inhibited as there is depletion of import factors. In contrast to this, plant importin α is strongly associated with nuclear membrane and is sufficient to mediate protein import independent of interaction with importin β.

But many times, the amino acids forming NLS are close together but not immediately adjacent to each other. e.g. NLS of nucleoplasmin (a protein involved in chromatin assembly) is bipartite, consisting of two parts – one part is a pair of lysine – arginine and second part is four lysines. Both these parts are 10 amino acids away from each other, but both are necessary for nuclear transport.

NLS are recognized by proteins which function as nuclear transport proteins. Most of the nuclear transport proteins are members of karyopherin protein family. They either function as importins (transporting molecules from cytoplasm to nucleus) or exportins (transporting molecules from nucleus to cytoplasm).

The Ran Cycle

The transport across nuclear membrane is facilitated by Ran cycle. Ran is a small GTP binding protein. The conformation and activity of Ran are regulated by binding of GTP and GTP hydrolysis. Ran shuttles between cytoplasm and nucleus depending on the accessory proteins present in each compartment.

RanGAP (Ran GTPase activating protein – enzyme which stimulates GTP hydrolysis to GDP) is localized on the cytoplasmic side of the NPC. So it facilitates hydrolysis of RanGTP to RanGDP outside the nucleus. Therefore the cytoplasm shows high concentration of RanGDP. Ran guanine nucleotide exchange factor (RanGEF - enzyme which stimulates conversion of GDP to GTP) is present inside the nucleus, so nucleus shows high amount of RanGTP, resulting in unequal distribution of RanGTP across the nuclear membrane. The concentration of RanGTP determines the directionality of the nuclear transport of cargo molecules. Ran regulates movement of cargo molecules by controlling the activity of nuclear transport receptors.

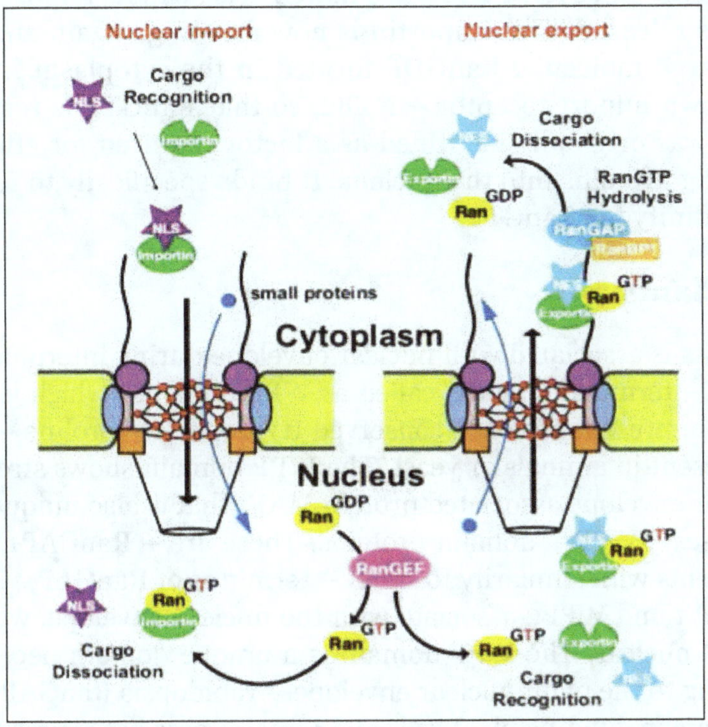

Receptor mediated nucleocytoplasmic transport – Import receptor recognizes and binds to NLScontaining cargos. The complex proceeds to nucleoplasm through centre of NPC. In the nucleus, the import complex encounters high RanGTP maintained by RanGEF. Binding of RanGTP to the importin releases the cargo, driving the transport reaction to the nucleus. Export receptor (NES) binds to its cargo in the nucleus in the presence of RanGTP and passes through the NPC to reach the cytoplasm. Here RanGAP1 and cofactor RanBP1 cause the complex to dissociate and RanGTP to hydrolyze, driving the transport reaction to the cytoplasm. RanGDP and empty export receptors are recycled back to nucleus; while inport receptor / RanGTP complexes are recycled back to the cytoplasm where they are dissociated by RanGAP1 and RanBP1.

In cytoplasm, importin α, β bind to the NLS present on the cargo molecule and forms a cargo – importin complex. Then this cargo – importin complex binds to specific proteins in the meshwork of nucleoporins. Transport is facilitated by sequential binding of this cargo – importin complex to specific nucleoporins located further and further, like a chain, towards the nucleus. Once the complex reaches the nucleus, then importin β from the cargo – importin complex tightly binds to RanGTP present in the nucleus. This binding causes change in importin conformation and thereby cargo molecule is dissociated from cargo – importin complex. As this complex dissociates, import of cargo molecule is irreversible. The importin – RanGTP complex then leaves the nucleus so that importins return to the cytoplasm. After reaching the cytoplasm, RanGTP complex undergoes hydrolysis which is facilitated by RanGAP, and becomes RanGDP. Once RanGDP is formed, importin can no longer bind to it and is released. This importin is now free to go again and form complex with another cargo molecule. RanGDP formed in the cytoplasm is transported to nucleus by its own import receptor – NTF2, so that RanGTP is regenerated in the nucleus. NTF2 was originally identified as a factor required for efficient import of NLS – containing proteins into the nucleus. It binds specifically to RanGDP and has no detectable affinity for RanGTP.

RanGAP in Plants

RanGAP in plants is associated with nuclear envelope during interphase and contains plant specific N – terminal domain, called as WPP domain, (which is a N – terminal targeting domain, present after the conserved tryptophan – proline – proline motif), which is not present in animals or yeast. The WPP domain shows strong similarity to the small nuclear envelope associated protein MAF1 that is also unique to plants. Arabidopsis expresses 5 WPP – domain proteins. These are – RanGAP1, RanGAP2, and three short proteins with similarity to the N – terminus of RanGAP, viz. WPP1, WPP2 and WPP3. WPP 1 and WPP2 associate with the nuclear envelope, whereas, WPP3 is cytoplasmic and nuclear. The WPP domain is a unique domain necessary and sufficient for targeting to the plant nuclear envelope. Arabidopsis RanGAP1 appears at the emerging cell plate in dividing plant cells and remains associated with its growing rim.

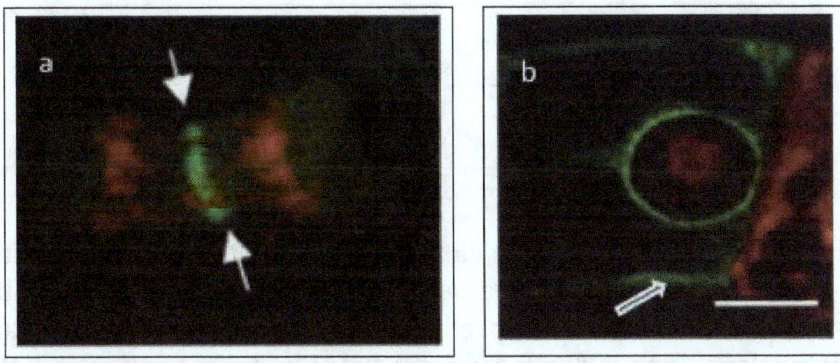

At RanGAP1- GFP in Arabidopsis – a) the arrows indicate association of AtRanGAP1 GFP with the position of the cell plate as early as anaphase. B) AtRanGAP1 – GFP observed as green fluorescence at the cell walls. Red staining is due to SYTO 82 orange for nucleic acids.

This plant specific mitotic targeting depends on the WPP domain and requires same amino acids as nuclear rim association. WPP domain interacting protein 1 (WIP1) was found to bind RanGAP as well as WPP domain. WIP1 –like proteins were identified from many plant species. In Arabidopsis, WIP2 and WIP3 encode most closely related proteins. These WIPs target RanGAP to the nuclear envelope. In addition to transport through nuclear pores, Ran cycle is crucially involved in assembly of NPC, in regulating spindle formation, kinetochore attachment and nuclear envelope assembly during mitosis.

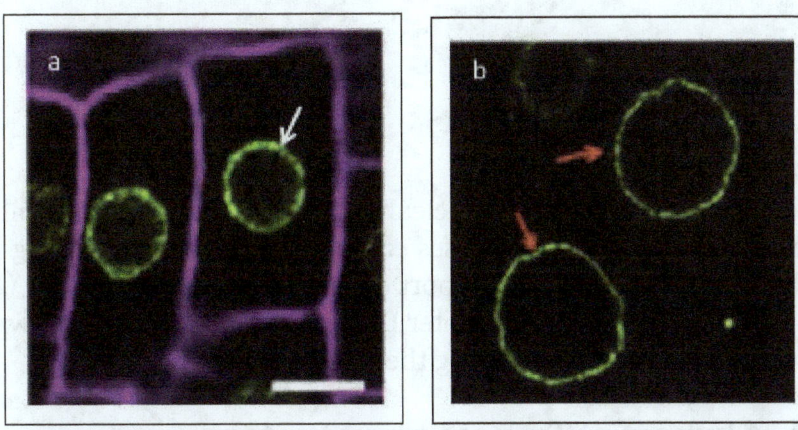

GFP-WIPs are targeted to the nuclear envelope (shown by white arrow) in Arabidopsis; 4b – RanGAP targeting to the nuclear envelope (but not the cell plate) depends on the WIP family.

Nuclear Import of RNA Polymerase II

RNA polymerase (Pol) II transcribes protein coding genes in the nucleus of eukaryotic cells. It is the conserved 12 – subunit enzyme. Till recently, it was not known how Pol II is imported into the nucleus. Pol II has 2 large subunits viz. Rpb1 and Rpb2 and 10 – 15 smaller subunits. As Pol II subunit sequences lack a known NLS, studies by Czeko et al (2011) have shown that Pol II likely binds to protein Iwr1 ("interacts with RNA polymerase II 1") wr1 is a 40 kDa protein with a bipartite NLS.

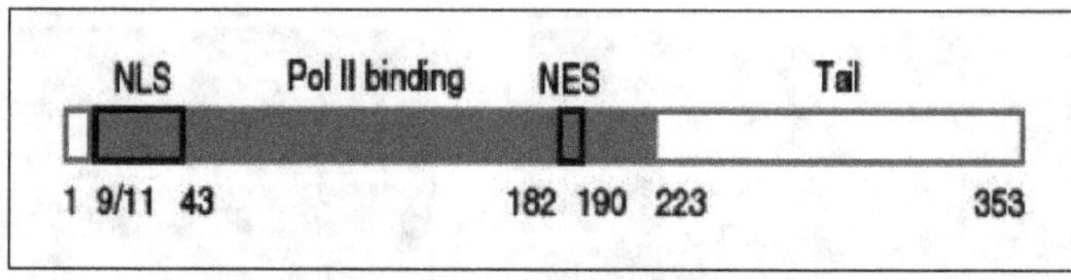

Schematic representation of yeast Iwr1.

Iwr1 binds Pol II in the active centre cleft between the two largest subunits, maybe facilitating or sensing complete Pol II assembly in the cytoplasm. Pol II is imported in the nucleus through N – terminal bipartite NLS which is recognized by karyopherin α. Once in the nucleus, Iwr1 is displaced from Pol II by transcription initiation factors and nucleic acids, enabling its export and recycling. Iwr1 function is Pol II specific and transcription independent.

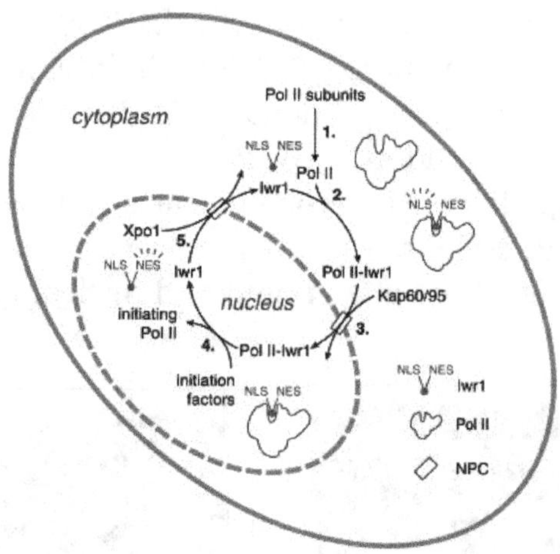

Cyclic model for Iwr1 directed Pol II nuclear transport – (1) Pol II assembles from its subunits in the cytoplasm; (2) Iwr1 senses complete assembly by binding in the Pol II cleft; (3) The Iwr1 NLS directs Pol II import through NPC via Kap60/95; (4) Assembly of Pol II with initiation factors on promoter DNA releases Iwr1; (5) The Iwr1 NES facilitates nuclear export via Xpo, completing the cycle.

Regulation of Import of Transcription Factors

The activities of nuclear proteins can be controlled during protein transport to the nucleus. E.g. transcription factors – they are functional only when they are present in the nucleus. So regulation of their import is a novel means of controlling gene expression. The regulated nuclear import of transcription factors and protein kinases play an important role in controlling cell behavior in response to environmental changes.

There are two mechanisms of regulation:

- In one mechanism, transcription factors or other proteins associate with cytoplasmic proteins that mask their NLS. As these signals are now not recognized, the proteins remain in cytoplasm. E.g. NF–κB. It is activated in response to variety of extracellular signals in mammalian cells. In unstimulated cells, NF–κB is found as an inactive complex due to binding of an inhibitory protein, IκB in the cytoplasm. This binding masks the NLS and so prevents its transport to the nucleus. In stimulated cells, IκB is phosphorylated and it is degraded by ubiquitin – mediated proteolysis and allows NF– κB to be transported to the nucleus.

- Nuclear import of other transcription factor is regulated directly by their phosphorylation rather than binding with inhibitory proteins. E.g. yeast transcription factor Pho4. It is phosphorylated at atypical serine residue adjacent to its NLS. This phosphorylation inhibits nuclear import of Pho4. Under appropriate conditions, regulated dephosphorylating of this site takes place which activates Pho4 and it is transported to nucleus.

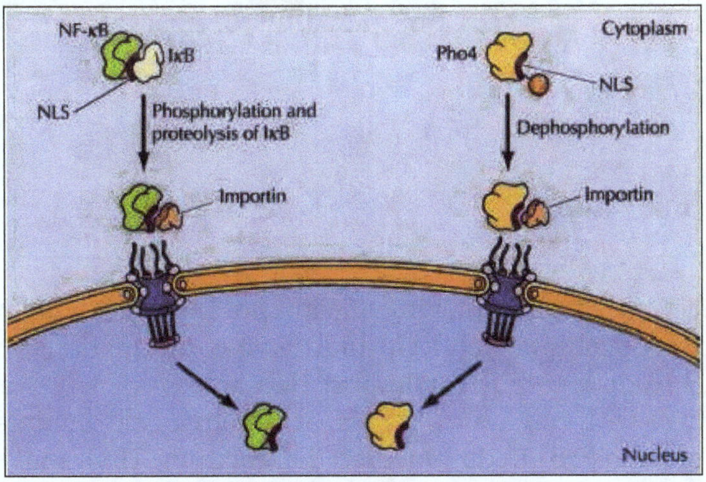

Regulation of nuclear import.

Nucleocytoplasmic Trafficking of Proteins

Nucleocytoplasmic trafficking of proteins is a highly selective process. Regulated nucleocytoplasmic transport is an important step in plant signaling pathways including hormone responses, responses to biotic and abiotic stresses, environmental changes and endogenous regulatory systems like circardian clocks. In plants, number of pathways is known involving a step of regulated nucleocytoplasmic partitioning, but none of these mechanisms are mutually exclusive. All these pathways are grouped into four categories.

Protein Modification or Conformational Change that either Masks or Exposes NLS or NES

Nuclear localization of phytochrome occurs through at least two distinct mechanisms:

- Light stable phytochrome B (phy B), a dimer, enters nucleus through light mediated change in conformation that unmasks NLS. In dark NLS exposure is restricted. After red light absorbance, structural changes occur, opening up the dimer and exposing NLS. In contrast to this simple unmasking of phy B NLS, phytochrome A (phy A) nuclear entry requires interaction with plant specific cytosolic factors, FHY1 and/or FHL. Both proteins (FHY1 and FHL) contain NLS and NES at their N-termini and phy A interaction domain at C-termini. These proteins are able to homodimerize and heterodimerize through their Ctermini and both are involved in trafficking phy A. The presence of both NLS and NES in these proteins suggests their shuttling between nucleus and cytosol.

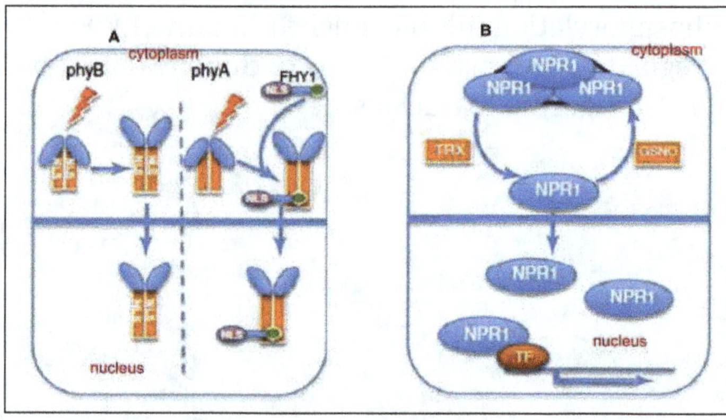

Regulated nucleocytoplasmic trafficking of plant proteins A) Phytochrome regulation by NLS exposure/association. Red light induces a conformational change in phyB, exposing an NLS leading to nuclear import (left side of panel). Red-light-induced association of FHY1 with phyA allows nuclear import of the complex via the FHY1 NLS (right side of panel). B) Redox-mediated regulation. Thioredoxin (TRX)- mediated reduction of NPR1 complexes allows import of NPR1 monomers to the nucleus. Monomer oxidation by S-nitrosglutathione (GSNO) promotes oligomerization and nuclear exclusion.

- Another mechanism of regulated partitioning is through reversible conformation change in defense gene expression, NPR1. Innate immune responses are many times associated with changes in cellular oxidative states. NPR1 localization responds to these redox changes. In non-challenged plants, NPR1 accumulates in cytosol and undergoes oligomerization via inter-molecular disulphide bond formation, which is facilitated by Snitrosylation (covalent attachment of NO to a reactive cysteine by S-nitrosglutathione – GSNO). Such oligomerization through inter-molecular disulphide bond formation is unusual in reductive environment of cytoplasm. After pathogen attack, salicylic acid concentration increases and this causes reduction of disulphide bonds and release of NPR1 monomers. This monomerization is facilitated by thioredoxins and the monomers are then localized in nucleus. The balance between these competing reactions can fine tune the nucleocytoplasmic partitioning of NPR1.

Nuclear Import by Heterodimerization or Heterocomplex Formation

It plays major role in plant pathogen interactions. This trafficking pathway is based on increasing number of examples of soluble immune receptors which initially were considered to be cytoplasmic, but are now found quite exclusively in the nucleus after pathogen attack. E.g. nuclear import of tobacco TIR-NB-LRR immune receptor N, which was previously shown to be triggered by viral effector p50. have shown that this interaction is mediated by a chlorophyll protein – N-receptor interacting protein (NRIP1). In uninfected plants, NRIP1 is a chlorophyll localized functional rhodanese sulphur transferase. In presence of TMV p50, NRIP1 localizes both in cytoplasm and nucleus. But precise role of nuclear NRIP 1 in signaling is not yet known.

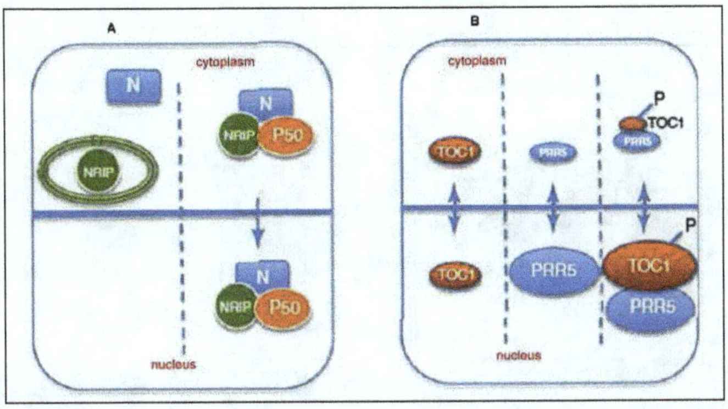

Proposed nuclear import by diversion from chloroplast targeting in the absence of the viral effector p50, NRIP is located in the chloroplast and the resistance protein N is located in the cytoplasm (left side of panel). In the presence of the viral effector p50, the chloroplast protein NRIP forms a complex with p50 and the resistance protein N in the cytoplasm, which enters the nucleus (right side of panel). B) Regulation by heterocomplex formation. Nucleocytoplasmic partitioning of circadian clock factor, TOC1 is shifted towards the nucleus in the presence of PRR5. Heterocomplexing of the two increases TOC1 phosphorylation, but has no detectable effect on PRR5 partitioning or phosphorylation state.

Similar mechanism has been identified in Arabidopsis circardian clock. It shows two closely related circardian factors – TOC1 and PRR5 (a part of pseudoresponse regulator proteins [PRR] family). TOC1 and PRR5 show peak expression during early evening and are able to heterodimerize in vivo. A large proportion of PRR5 is present in nucleus at all times, independent of TOC1 levels. But when PRR5 is over expressed, TOC1 is concentrated in nucleus. Both proteins (TOC1 & PRR5) possess C-terminal NLS. C-terminal deleted TOC1 (N-TOC1) is exclusively found in cytoplasm. In presence of PRR5, N-TOC1 is localized in nucleus and a hyper-phosphorylated form appears in both cytoplasm and nucleus. So, Prr5 facilitates both nuclear accumulation and hyperphosphorylation of TOC1.

Nuclear or Cytoplasmic Retention by Protein-protein Interactions and its Release by Protein Modification

An interesting alternative to enhanced nuclear import is increased nuclear retention through heterocomplex based inhibition of nuclear export. This is the emerging model of partitioning control of photomorphogenesis through COP1. In dark, COP1 is concentrated in the nucleus, where it inhibits photomorphogenic development through proteolysis of light regulatory transcription factors. In light, COP1 is translocated to cytosol and gets inactivated. E.g. in potato immune receptor – Rx1, cytoplasmic retention of Rx1 confers extreme resistance to potato virus X through interaction with viral coat protein.

Cytoplasmic Retention by Membrane Association and its Release by Proteolysis

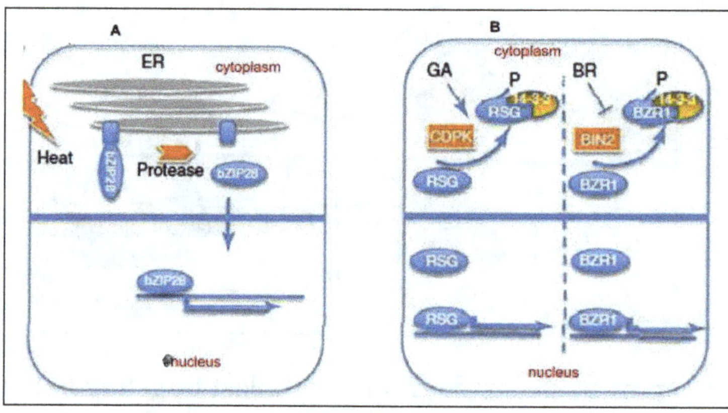

In Arabidopsis, a large number of membrane bound transcription factors (anchored either to endoplasmic reticulum or plasma membrane) are translocated to nucleus based on a stimulus. E.g. a group of bZIP proteins – Expression of bZIP28 in Arabidopsis is upregulated by heat. Heat stress induces proteolytic release of bZIP28 from membrane of endoplasmic reticulum and its redistribution to nucleus. Likewise, antibiotics or salt stress also can induce proteolytic cleavage. Similar release mechanisms have been observed for plasma membrane associated proteins. E.g. the transcription factor NTL6 is proteolytically released from plasma membrane by cold. It then enters nucleus and activates a subset of pathogen response genes.

Another example of Cytoplasmic retention is by 14-3-3 binding. GA3 encodes ent-kaurene oxidase that acts early in gibberellic acid biosynthetic pathway. A transcription factor – RSG (repression of shoot growth) activates GA3. In absence of gibberellic acid, RSG distributes equally between cytoplasm and nucleus. When gibberellic acid is applied, RSG is rapidly partitioned in cytoplasm through phosphorylation and subsequent binding to a 14-3-3 protein. This response results in gibberellic acid mediated down regulation of its own biosynthesis.

Similar mechanism is observed in brassinosteroid signaling. Phosphorylation of transcription factor – brassinazole resistant 1 (BZR1) promotes binding to 14-3-3 protein, causing Cytoplasmic retention and/or nuclear export.

Regulation by proteolytic release from membrane association, Heat stress activates proteolytic cleavage of the ER-associated transcription factor bZIP28. B) Regulation by 14-3-3 interaction. GA initiates sequestration of RSG in the cytoplasm through CDPK1-dependent phosphorylation and subsequent binding to a 14-3-3 protein (left side of panel). BR initiates events that lead to dephosphorylated BZR1 accumulation in the nucleus. Absent BR, BIN2 phosphorylation of BZR1 sequesters it in the cytoplasm through 14-3-3 binding,followed by proteolysis (right side of panel).

Nuclear Export

Transport across the NPC is very much essential to exchange materials as well as information in a living cell. Initially this transport was thought to be important only because translation occurs in cytoplasm and transcription in nucleus. But now we know that nuclear export of proteins is very essential for regulation of specific cellular processes and for signalling.

Nuclear Export Receptors

There are two different classes of nuclear export receptors:

- Members of this class bind to Ran, so they are also called as Ran binding proteins (RanBPs).
- Members of this class do not bind to Ran and are structurally unrelated to RanBPs.

RanBPs form a family of related proteins called as importin β. Members of this family, those which facilitate cargo import are called as importins or transportins and those which facilitate export are called as exportins. There are 17 genes encoding importin β – like nuclear transport receptors in Arabidopsis, whereas, human genome contains at least 21 genes. These nuclear transport receptors are well conserved between species. Many plant nuclear transport receptors are known e.g. Xpo1a, Xpo1b, Xpo-t, RanGAP1 and NTF2a.

RanGTP gradient over nuclear envelope drives nuclear export. Xpo1a binds directly to many substrates containing leucine – rich nuclear export signal (NES) along with RanGTP to form a export complex. This complex associates with the NPC. Xpo1a interacts with the FG nucleoporins and the export complex is transported to cytoplasm. In cytoplasm there are two proteins – RanBP1 (Ran binding protein 1) and RanGAP (ran GTPase activating protein) which together bring about hydrolysis of GTP from RanGTP. As a result, the export complex dissociates and the export cargo is set free in cytoplasm. RanGTP hydrolyses to RanGDP. Xpo1a has very low affinity to cargo molecules in absence of RanGTP, so Xpo1a also dissociates from the export complex and it recycles back to nucleus on its own, as it has ability to bind to FG nucleoporins.

Some export cargoes like pre-60S ribosomal subunit requires an adapter protein, through which it binds to the export receptor Xpo1a. Once the pre-60S ribosomal subunit reaches cytoplasm, the adapter is recycled back to the nucleus.

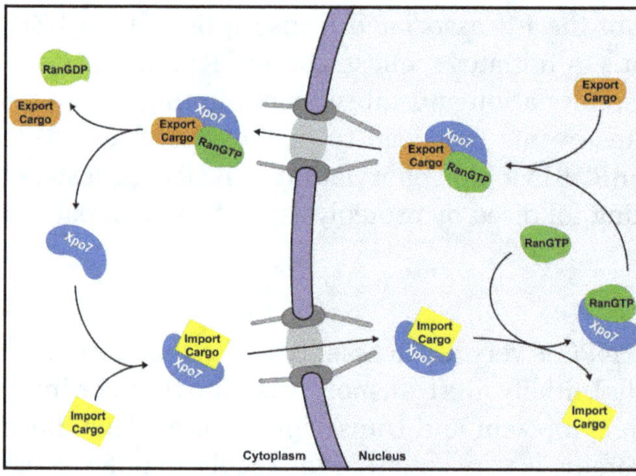

Nuclear export by importin β-like receptors - Xpo1 (exportin 1) binds directly to many substrates containing a leucine-rich NES, co-operatively with RanGTP that is abundant in the nucleus. This export complex docks to the NPC and is translocated to the Cytoplasmic side via direct interactions of Xpo1 with FG nucleoporins. Here the export complex encounters two Cytoplasmic regulators (RanBP1 and RanGAP) of the RanGTPase cycle, that act together to hydrolyze GTP on Ran. As a consequence, the export complex disaasembles and the export cargo is released into the cytoplasm. As RanGTP concentrations are very low in cytoplasm, and because Xpo1 shows very low affinity to cargos in absnce of RanGTP, Xpo1 recycles back to the nucleus on its own due to its ability to interact with FG nucleoporins. Some export cargos, like pre-60S ribosomal subunit need an adapter protein (Adapt) that bridges its interaction with the export receptor. The export adapter is then recycled to the nucleus by an import receptor.

Plants produce one more exportin – exportin 2, also known as importin α re-exporter, or CAS (cellular apoptosis susceptibility protein). Importin α proteins are import adapters, which bind to many karyophilic proteins, contain a basic NLS and link them to import receptor. Importin α are released in the nucleus which has to be recycled into the cytoplasm by an export receptor, as it cannot cross the NPC on its own. Importin α is recycled by this CAS.

All proteins that are imported into the nucleus are accumulated in the nucleus, but some of these proteins have to shuttle continuously between the nucleus and cytoplasm. Members of importin β family are able to cross the NPC on their own as they can interact with FG nucleoporins. But all other proteins require a nuclear export pathway to leave the nucleus. Importin α requires CAS nuclear export pathway, for which the receptor has been characterized in plants. Nuclear export pathways are used to ensure distinct localization of specific proteins and restrict them to the cytoplasm.

Nuclear RNA Export

Eukaryotic cells export several different classes of RNA molecules from the nucleus, where they are transcribed, to the cytoplasm where majority of them participate in different aspects of protein synthesis. For all classes of RNA molecules, nuclear export depends on the assembly of RNA into appropriate rib nucleoprotein complexes. Hence, nuclear export also appears to function as an important proof-reading mechanism.

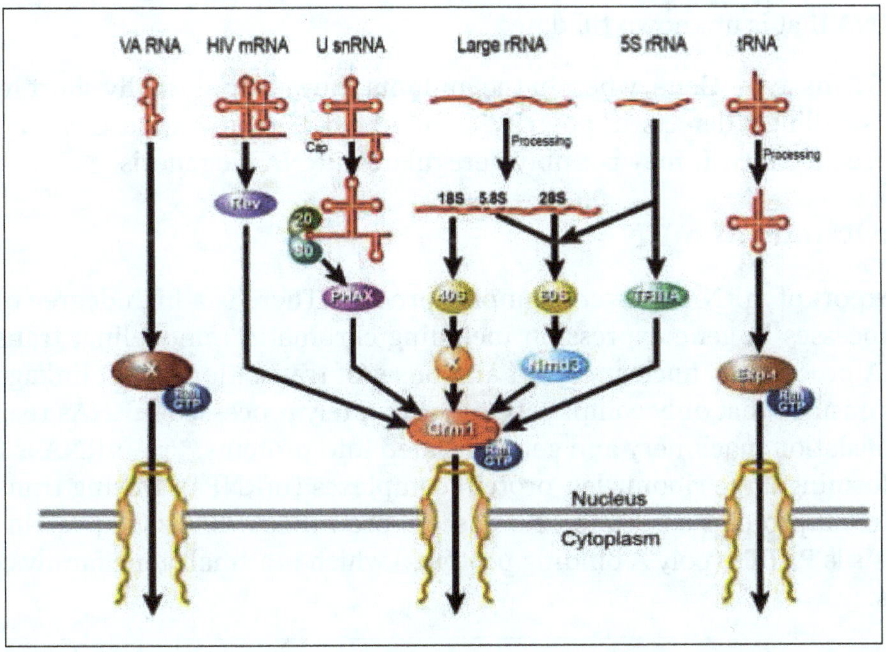

Karyopherin – mediated nuclear RNA export pathways – An overview of the key factors involved in mediating the nuclear export of different classes of non coding RNA, as well as the minor class of mRNAs. In addition to late HIV mRNAs, a small number of cellular RNAs may also be exported from the nucleus in a Crm1-dependent manner. The nuclear export pathway used by VA RNAs is also used by other small, non coding RNAs including Y RNAs and possibly, pre-miRNAs.

The bulk of cellular nucleocytoplasmic transport is mediated by factors that belong to a single family of nuclear transport receptors called as karyopherins or importins/exportins. These karyopherins depend on the biological activity of an important cofactor – GTPase Ran. Interference with Ran GTP gradient across the nuclear membrane inhibits nuclear export of most RNAs, except bulk mRNA export. This bulk mRNA export is not dependent on Ran, but is mediated by nuclear export factors that are not members of karyopherin family of proteins. Several different types of RNAs are exported from the nucleus, like large rRNAs, mRNAs, U snRNAs, tRNAs and 5S rRNAs. These RNAs largely use distinct pathways to exit the nucleus.

Export of Small RNAs

Two genes encoding importin β – like nuclear transport receptors were characterized in Arabidopsis mutant:

- PSD (paused) – Encodes export receptor of tRNAs. PSD mutants showed affected meristem initiation during embryogenesis, delayed transition to reproductive development, delayed root growth and lateral root initiation. But the plants are viable, indicating a possible existence of another nuclear export pathway for tRNA that is unknown till date.
- HST (hasty) – Gene, wherein the mutants showed abnormally short hypocotyls in seedlings, defects in polarity, accelerated vegetative phase change and reduced fertility. It may possibly deregulate miRNA biogenesis.

Nuclear Export of mRNA

Nuclear export of mRNA is a very complex process. There is a high degree of integration of processes in gene expression including chromatin remodeling, transcription, pre-mRNA processing, nuclear export and onset of translation. Tight linkage of these processes ensures that only completely and accurately processed mRNAs reach the cytosolic translation machinery and get translated into proteins. Pre-mRNA is bound by proteins forming large ribonucleo-protein complexes (mRNPs). During transcription, many different proteins associate with nascent pre-mRNA. One such protein family in Arabidopsis is PABPs (poly A binding proteins) which is a small gene family consisting of 8 genes.

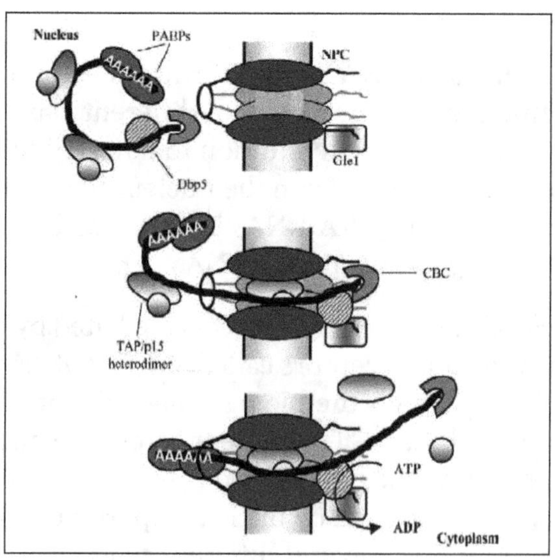

Figure shows, nuclear export of mRNA: Simplified model to highlight plant proteins that may be implicated in this pathway and to indicate a possible mechanism that is discussed to create directionality in mRNA export. mRNA associates in the nucleus with cap-binding complex (CBC) consisting of CBC80 and CBC20 (single genes each in Arabidopsis), with poly(A) binding proteins (PABPs; multigene family in Arabidopsis containing eight members), with Dbp5 (candidate homolog in Arabidopsis : LOS4), and with the RNA export receptor TAP/p15 (TAP is missing in Arabidopsis ; the candidate homolog of p15 in Arabidopsis is NTL). Once the mRNP-associated Dbp5 has reached the cytoplasmic side of the NPC, it associates with cytoplasmic nucleoporins of the NPC and with Gle1, which in turn activates the ATP dependent helicase of the Dbp5. This may result in the removal of RNA-binding proteins, among them the TAP/p15 heterodimer, thereby preventing the return of the transport complex to the nucleus. Many other proteins and protein complexes associate with mRNPs in the nucleus and/or in the cytoplasm.

After capping of 5' end of mRNA, cap binding complex (CBC) is formed at cap structure and it accompanies mRNAs to cytoplasm. Arabidopsis contains two single genes that encode large and small subunit of CBC – CBC 80 and CBC 20 respectively. CBC 80 is mainly localized in nucleus at steady state, but small amount is also seen in cytoplasm, indicating its shuttling. mRNA associates with CBC, PABPs, Dbp5 (nucleo-cytoplasmic shuttle protein) and RNA export receptor TAP/p15. The mRNA contacts the NPC via nucleoporins located in the nuclear basket. Once the mRNA associated Dbp5 has reached the cytoplasmic side of the NPC, it associates with cytoplasmic nucleoporins and with Gle1 (RNA export factor), which in turn activates ATP dependent helicase of Dbp5. This results in removal of RNA binding proteins - TAP/p15 and so preventing the return of transport complex back to the nucleus.

Chapter 4

Plant Cell Wall

Plant cell wall is a structure present around the plasma membrane of plant cells. It is a rigid structure and provides tensile strength to the cell, and protects against mechanical and osmotic stress. Plant cell walls are mainly composed of cellulose, which is a linear polymer of glucose. This chapter discusses in detail the concepts related to plant cell wall.

Cell Wall

A cell wall is a structural layer surrounding some types of cells, situated outside the cell membrane. It can be tough, flexible, and sometimes rigid. It provides the cell with both structural support and protection, and also acts as a filtering mechanism. Cell walls are present in most prokaryotes (except mycoplasma bacteria), in algae, plants and fungi but rarely in other eukaryotes including animals. A major function is to act as pressure vessels, preventing over-expansion of the cell when water enters.

The composition of cell walls varies between species and may depend on cell type and developmental stage. The primary cell wall of land plants is composed of the polysaccharides cellulose, hemicellulose and pectin. Often, other polymers such as lignin, suberin or cutin are anchored to or embedded in plant cell walls. Algae possess cell walls made of glycoproteins and polysaccharides such as carrageenan and agar that are absent from land plants. In bacteria, the cell wall is composed of peptidoglycan. The cell walls of archaea have various compositions, and may be formed of glycoproteinS-layers, pseudopeptidoglycan, or polysaccharides. Fungi possess cell walls made of the glucosamine polymer chitin. Unusually, diatoms have a cell wall composed of biogenic silica.

History

A plant cell wall was first observed and named (simply as a "wall") by Robert Hooke in 1665. However, "the dead excrusion product of the living protoplast" was forgotten, for almost three centuries, being the subject of scientific interest mainly as a resource for industrial processing or in relation to animal or human health.

In 1804, Karl Rudolphi and J.H.F. Link proved that cells had independent cell walls. Before, it had been thought that cells shared walls and that fluid passed between them this way.

The mode of formation of the cell wall was controversial in the 19th century. Hugo von Mohl (1853, 1858) advocated the idea that the cell wall grows by apposition. Carl Nägeli (1858, 1862, 1863) believed that the growth of the wall in thickness and in area was due to a process termed intussusception. Each theory was improved in the following decades: the apposition (or lamination) theory by Eduard Strasburger (1882, 1889), and the intussusception theory by Julius Wiesner (1886).

In 1930, Ernst Münch coined the term *apoplast* in order to separate the "living" symplast from the "dead" plant region, the latter of which included the cell wall.

By the 1980s, some authors suggested replacing the term "cell wall", particularly as it was used for plants, with the more precise term "extracellular matrix", as used for animal cells, but others preferred the older term.

Properties

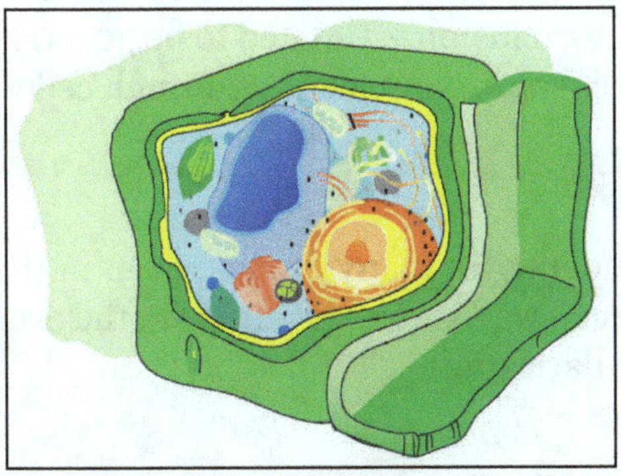

Diagram of the plant cell, with the cell wall in green.

Cell walls serve similar purposes in those organisms that possess them. They may give cells rigidity and strength, offering protection against mechanical stress. In multicellular organisms, they permit the organism to build and hold a definite shape (morphogenesis). Cell walls also limit the entry of large molecules that may be toxic to the cell. They further permit the creation of stable osmotic environments by preventing osmotic lysis and helping to retain water. Their composition, properties, and form may change during the cell cycle and depend on growth conditions.

Rigidity of Cell Walls

In most cells, the cell wall is flexible, meaning that it will bend rather than holding a fixed shape, but has considerable tensile strength. The apparent rigidity of primary plant tissues is enabled by cell walls, but is not due to the walls' stiffness. Hydraulic turgor pressure creates this rigidity, along with the wall structure. The flexibility of the cell walls is seen when plants wilt, so that the stems and leaves begin to droop, or in seaweeds that bend in water currents. As John Howland explains:

Think of the cell wall as a wicker basket in which a balloon has been inflated so that it exerts pressure from the inside. Such a basket is very rigid and resistant to mechanical damage. Thus does the prokaryote cell (and eukaryotic cell that possesses a cell wall) gain strength from a flexible plasma membrane pressing against a rigid cell wall.

The apparent rigidity of the cell wall thus results from inflation of the cell contained within. This inflation is a result of the passive uptake of water.

In plants, a secondary cell wall is a thicker additional layer of cellulose which increases wall rigidity. Additional layers may be formed by lignin in xylem cell walls, or suberin in cork cell walls. These compounds are rigid and waterproof, making the secondary wall stiff. Both wood and bark cells of trees have secondary walls. Other parts of plants such as the leaf stalk may acquire similar reinforcement to resist the strain of physical forces.

Permeability

The primary cell wall of most plant cells is freely permeable to small molecules including small proteins, with size exclusion estimated to be 30-60 kDa. The pH is an important factor governing the transport of molecules through cell walls.

Evolution

Cell walls evolved independently in many groups, even in the photosynthetic eukaryotes. In these lineages, the cell wall is closely related to the evolution of multicellularity, terrestrialization and vascularization.

Plant Cell Walls

The walls of plant cells must have sufficient tensile strength to withstand internal osmotic pressures of several times atmospheric pressure that result from the difference in solute concentration between the cell interior and external solutions. Plant cell walls vary from 0.1 to several μm in thickness.

Layers

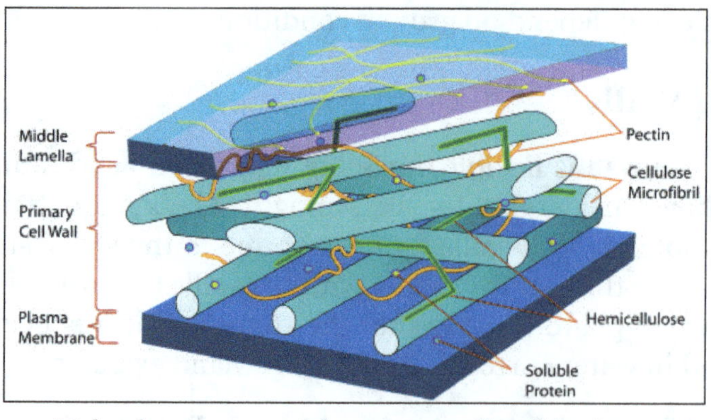

Molecular structure of the primary cell wall in plants.

Up to three strata or layers may be found in plant cell walls:

- The primary cell wall, generally a thin, flexible and extensible layer formed while the cell is growing.
- The secondary cell wall, a thick layer formed inside the primary cell wall after the cell is fully grown. It is not found in all cell types. Some cells, such as the conducting cells in xylem, possess a secondary wall containing lignin, which strengthens and waterproofs the wall.
- The middle lamella, a layer rich in pectins. This outermost layer forms the interface between adjacent plant cells and glues them together.

Composition

In the primary (growing) plant cell wall, the major carbohydrates are cellulose, hemicellulose and pectin. The cellulose microfibrils are linked via hemicellulosic tethers to form the cellulose-hemicellulose network, which is embedded in the pectin matrix. The most common hemicellulose in the primary cell wall is xyloglucan. In grass cell walls, xyloglucan and pectin are reduced in abundance and partially replaced by glucuronarabinoxylan, another type of hemicellulose. Primary cell walls characteristically extend (grow) by a mechanism called acid growth, which involves turgor-driven movement of the strong cellulose microfibrils within the weaker hemicellulose/pectin matrix, catalyzed by expansin proteins. The outer part of the primary cell wall of the plant epidermis is usually impregnated with cutin and wax, forming a permeability barrier known as the plant cuticle.

Secondary cell walls contain a wide range of additional compounds that modify their mechanical properties and permeability. The major polymers that make up wood (largely secondary cell walls) include:

- Cellulose, 35-50%.
- Xylan, 20-35%, a type of hemicellulose.
- Lignin, 10-25%, a complex phenolic polymer that penetrates the spaces in the cell wall between cellulose, hemicellulose and pectin components, driving out water and strengthening the wall.

Additionally, structural proteins (1-5%) are found in most plant cell walls; they are classified as hydroxyproline-rich glycoproteins (HRGP), arabinogalactan proteins (AGP), glycine-rich proteins (GRPs), and proline-rich proteins (PRPs). Each class of glycoprotein is defined by a characteristic, highly repetitive protein sequence. Most are glycosylated, contain hydroxyproline (Hyp) and become cross-linked in the cell wall. These proteins are often concentrated in specialized cells and in cell corners. Cell walls of the epidermis may contain cutin. The Casparian strip in the endodermis roots and cork cells of plant bark contain suberin. Both cutin and suberin are polyesters that function as permeability barriers to the movement of water. The relative composition of carbohydrates, secondary compounds and proteins varies between plants and between the cell type and age. Plant cells walls also contain numerous enzymes, such as hydrolases, esterases, peroxidases, and transglycosylases, that cut, trim and cross-link wall polymers.

Secondary walls - especially in grasses - may also contain microscopic silica crystals, which may strengthen the wall and protect it from herbivores.

Cell walls in some plant tissues also function as storage deposits for carbohydrates that can be broken down and resorbed to supply the metabolic and growth needs of the plant. For example, endosperm cell walls in the seeds of cereal grasses, nasturtium and other species, are rich in glucans and other polysaccharides that are readily digested by enzymes during seed germination to form simple sugars that nourish the growing embryo.

Formation

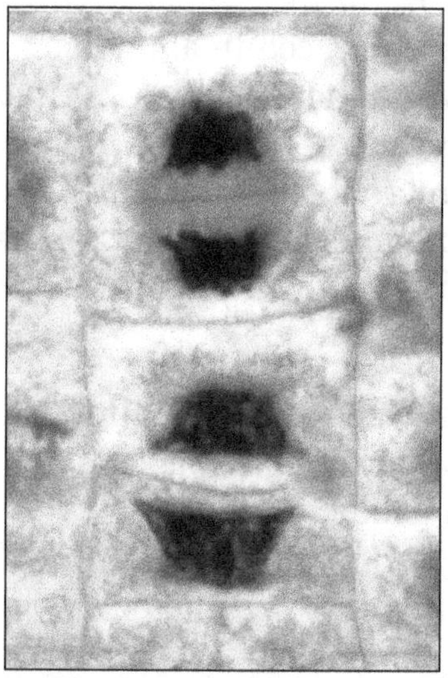

Photomicrograph of onion root cells, showing the centrifugal development of new cell walls (phragmoplast).

The middle lamella is laid down first, formed from the cell plate during cytokinesis, and the primary cell wall is then deposited inside the middle lamella. The actual structure of the cell wall is not clearly defined and several models exist - the covalently linked cross model, the tether model, the diffuse layer model and the stratified layer model. However, the primary cell wall, can be defined as composed of cellulosemicrofibrils aligned at all angles. Cellulose microfibrils are produced at the plasma membrane by the cellulose synthase complex, which is proposed to be made of a hexameric rosette that contains three cellulose synthase catalytic subunits for each of the six units. Microfibrils are held together by hydrogen bonds to provide a high tensile strength. The cells are held together and share the gelatinous membrane called the *middle lamella*, which contains magnesium and calciumpectates (salts of pectic acid). Cells interact though plasmodesmata, which are inter-connecting channels of cytoplasm that connect to the protoplasts of adjacent cells across the cell wall.

In some plants and cell types, after a maximum size or point in development has been reached, a *secondary wall* is constructed between the plasma membrane and primary wall. Unlike the primary wall, the cellulose microfibrils are aligned parallel in layers, the orientation changing slightly with each additional layer so that the structure becomes helicoidal. Cells with secondary cell walls can be rigid, as in the gritty sclereid cells in pear and quince fruit. Cell to cell communication is possible through pits in the secondary cell wall that allow plasmodesmata to connect cells through the secondary cell walls.

Fungal Cell Walls

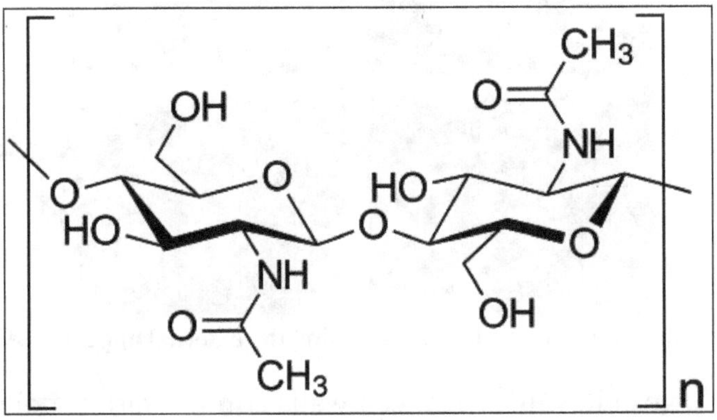

Chemical structure of a unit from a chitin polymer chain

There are several groups of organisms that have been called "fungi". Some of these groups (Oomycete and Myxogastria) have been transferred out of the Kingdom Fungi, in part because of fundamental biochemical differences in the composition of the cell wall. Most true fungi have a cell wall consisting largely of chitin and other polysaccharides. True fungi do not have cellulose in their cell walls.

True Fungi

In fungi, the cell wall is the outer-most layer, external to the plasma membrane. The fungal cell wall is a matrix of three main components:

- chitin: polymers consisting mainly of unbranched chains of β-(1,4)-l inked-N-Acetylglucosamine in the Ascomycota and Basidiomycota, or poly-β-(1,4)-linked-N-Acetylglucosamine (chitosan) in the Zygomycota. Both chitin and chitosan are synthesized and extruded at the plasma membrane.
- glucans: glucose polymers that function to cross-link chitin or chitosan polymers. β-glucans are glucose molecules linked via β-(1,3)- or β-(1,6)- bonds and

provide rigidity to the cell wall while α-glucans are defined by α-(1,3)- and/or α-(1,4) bonds and function as part of the matrix.

- proteins: enzymes necessary for cell wall synthesis and lysis in addition to structural proteins are all present in the cell wall. Most of the structural proteins found in the cell wall are glycosylated and contain mannose, thus these proteins are called mannoproteins or mannans.

Other Eukaryotic Cell Walls

Algae

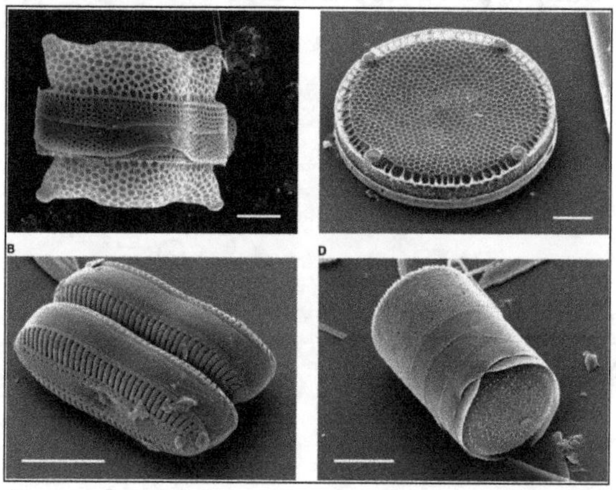

Scanning electronmicrographs of diatoms showing the external appearance of the cell wall

Like plants, algae have cell walls. Algal cell walls contain either polysaccharides (such as cellulose (a glucan)) or a variety of glycoproteins (Volvocales) or both. The inclusion of additional polysaccharides in algal cells walls is used as a feature for algal taxonomy.

- Mannans: They form microfibrils in the cell walls of a number of marine green algae including those from the genera, *Codium*, *Dasycladus*, and *Acetabularia* as well as in the walls of some red algae, like *Porphyra* and *Bangia*.
- Xylans.
- Alginic acid: It is a common polysaccharide in the cell walls of brown algae.
- Sulfonated polysaccharides: They occur in the cell walls of most algae; those common in red algae include agarose, carrageenan, porphyran, furcelleran and funoran.

Other compounds that may accumulate in algal cell walls include sporopollenin and calciumions.

The group of algae known as the diatoms synthesize their cell walls (also known as frustules or valves) from silicic acid (specifically orthosilicic acid, H_4SiO_4). The acid is polymerised intra-cellularly, then the wall is extruded to protect the cell. Significantly, relative to the organic cell walls produced by other groups, silica frustules require less energy to synthesize (approximately 8%), potentially a major saving on the overall cell energy budget and possibly an explanation for higher growth rates in diatoms.

In brown algae, phlorotannins may be a constituent of the cell walls.

Water Molds

The group Oomycetes, also known as water molds, are saprotrophicplant pathogens like fungi. Until recently they were widely believed to be fungi, but structural and molecular evidence has led to their reclassification as heterokonts, related to autotrophicbrown algae and diatoms. Unlike fungi, oomycetes typically possess cell walls of cellulose and glucans rather than chitin, although some genera (such as *Achlya* and *Saprolegnia*) do have chitin in their walls. The fraction of cellulose in the walls is no more than 4 to 20%, far less than the fraction of glucans. Oomycete cell walls also contain the amino acidhydroxyproline, which is not found in fungal cell walls.

Slime Molds

The dictyostelids are another group formerly classified among the fungi. They are slime molds that feed as unicellular amoebae, but aggregate into a reproductive stalk and sporangium under certain conditions. Cells of the reproductive stalk, as well as the spores formed at the apex, possess a cellulose wall. The spore wall has been shown to possess three layers, the middle of which is composed primarily of cellulose, and the innermost is sensitive to cellulase and pronase.

Prokaryotic Cell Walls

Bacterial Cell Walls

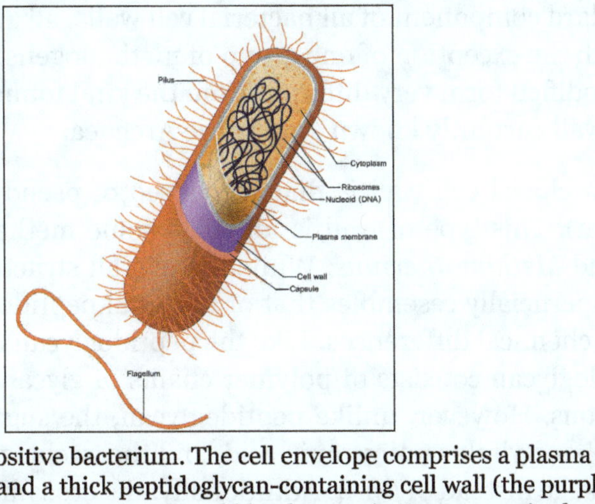

Diagram of a typical Gram-positive bacterium. The cell envelope comprises a plasma membrane, seen here in light brown, and a thick peptidoglycan-containing cell wall (the purple layer). No outer lipid membrane is present, as would be the case in Gram-negative bacteria. The red layer, known as the capsule, is distinct from the cell envelope.

Around the outside of the cell membrane is the bacterial cell wall. Bacterial cell walls are made of peptidoglycan (also called murein), which is made from polysaccharide chains cross-linked by unusual peptides containing D-amino acids. Bacterial cell walls are different from the cell walls of plants and fungi which are made of cellulose and chitin, respectively. The cell wall of bacteria is also distinct from that of Archaea, which do not contain peptidoglycan. The cell wall is essential to the survival of many bacteria, although L-form bacteria can be produced in the laboratory that lack a cell wall. The antibiotic penicillin is able to kill bacteria by preventing the cross-linking of peptidoglycan and this causes the cell wall to weaken and lyse. The lysozyme enzyme can also damage bacterial cell walls.

There are broadly speaking two different types of cell wall in bacteria, called Gram-positive and Gram-negative. The names originate from the reaction of cells to the Gram stain, a test long-employed for the classification of bacterial species.

Gram-positive bacteria possess a thick cell wall containing many layers of peptidoglycan and teichoic acids. In contrast, Gram-negative bacteria have a relatively thin cell wall consisting of a few layers of peptidoglycan surrounded by a second lipid membrane containing lipopolysaccharides and lipoproteins. Most bacteria have the Gram-negative cell wall and only the Firmicutes and Actinobacteria (previously known as the low G+C and high G+C Gram-positive bacteria, respectively) have the alternative Gram-positive arrangement. These differences in structure can produce differences in antibiotic susceptibility, for instance vancomycin can kill only Gram-positive bacteria and is ineffective against Gram-negative pathogens, such as *Haemophilus influenzae* or *Pseudomonas aeruginosa*.

Archaeal Cell Walls

Although not truly unique, the cell walls of Archaea are unusual. Whereas peptidoglycan is a standard component of all bacterial cell walls, all archaeal cell walls lack peptidoglycan, with the exception of one group of methanogens. In that group, the peptidoglycan is a modified form very different from the kind found in bacteria. There are four types of cell wall currently known among the Archaea.

One type of archaeal cell wall is that composed of pseudopeptidoglycan (also called pseudomurein). This type of wall is found in some methanogens, such as *Methanobacterium* and *Methanothermus*. While the overall structure of archaeal *pseudo*peptidoglycan superficially resembles that of bacterial peptidoglycan, there are a number of significant chemical differences. Like the peptidoglycan found in bacterial cell walls, pseudopeptidoglycan consists of polymer chains of glycan cross-linked by short peptide connections. However, unlike peptidoglycan, the sugar N-acetylmuramic acid is replaced by N-acetyltalosaminuronic acid, and the two sugars are bonded with a β,1-3 glycosidic linkage instead of β,1-4. Additionally, the cross-linking peptides are L-amino acids rather than D-amino acids as they are in bacteria.

A second type of archaeal cell wall is found in *Methanosarcina* and *Halococcus*. This type of cell wall is composed entirely of a thick layer of polysaccharides, which may be sulfated in the case of *Halococcus*. Structure in this type of wall is complex and not fully investigated.

A third type of wall among the Archaea consists of glycoprotein, and occurs in the hyperthermophiles, *Halobacterium*, and some methanogens. In *Halobacterium*, the proteins in the wall have a high content of acidicamino acids, giving the wall an overall negative charge. The result is an unstable structure that is stabilized by the presence of large quantities of positive sodiumions that neutralize the charge. Consequently, *Halobacterium* thrives only under conditions with high salinity.

In other Archaea, such as *Methanomicrobium* and *Desulfurococcus*, the wall may be composed only of surface-layer proteins, known as an *S-layer*. S-layers are common in bacteria, where they serve as either the sole cell-wall component or an outer layer in conjunction with polysaccharides. Most Archaea are Gram-negative, though at least one Gram-positive member is known.

Other Cell Coverings

Many protists and bacteria produce other cell surface structures apart from cell walls, external (extracellular matrix) or internal. Many algae have a sheath or envelope of mucilage outside the cell made of exopolysaccharides. Diatoms build a frustule from silica extracted from the surrounding water; radiolarians, foraminiferans, testate amoebae and silicoflagellates also produce a skeleton from minerals, called test in some groups. Many green algae, such as *Halimeda* and the Dasycladales, and some red algae, the Corallinales, encase their cells in a secreted skeleton of calcium carbonate. In each case, the wall is rigid and essentially inorganic. It is the non-living component of cell. Some golden algae, ciliates and choanoflagellatesproduces a shell-like protective outer covering called lorica. Some dinoflagellates have a theca of cellulose plates, and coccolithophorids have coccoliths.

An extracellular matrix is also present in metazoans. Its composition varies between cells, but collagens are the most abundant protein in the ECM.

Secondary Cell Wall

The secondary cell wall is a structure found in many plant cells, located between the primary cell wall and the plasma membrane. The cell starts producing the secondary cell wall after the primary cell wall is complete and the cell has stopped expanding.

Secondary cell walls provide additional protection to cells and rigidity and strength to the larger plant. These walls are constructed of layered sheaths of cellulose microfibrils, wherein the fibers are in parallel within each layer. The inclusion of lignin makes the secondary cell wall less flexible and less permeable to water than the primary cell wall. In addition to making the walls more resistant to degradation, the hydrophobic nature of lignin within these tissues is essential for containing water within the vascular tissues that carry it throughout the plant.

The secondary cell wall consists primarily of cellulose, along with other polysaccharides, lignin, and glycoprotein. It sometimes consists of three distinct layers - S_1, S_2 and S_3 - where the direction of the cellulose microfibrils differs between the layers.

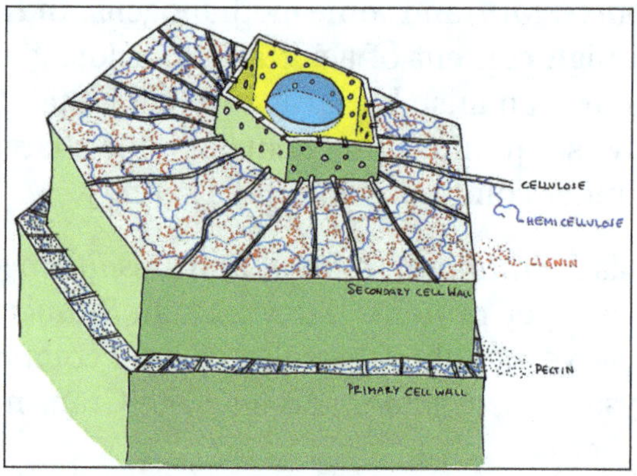

Plant cell overview, showing secondary cell wall.

The secondary cell wall has different ratios of constituents compared to the primary wall. An example of this is that secondary wall in wood contain polysaccharides called xylan, whereas the primary wall contains the polysaccharide xyloglucan. The cellulose fraction in secondary walls is also higher.Pectins may also be absent from the secondary wall, and unlike primary walls, no structural proteins or enzymes have been identified. Because of the low permeability through the secondary cell wall, cellular transport is carried out through openings in the wall called pits.

Wood consists mostly of secondary cell wall, and holds the plant up against gravity.

Some secondary cell walls store nutrients, such as those in the cotyledons and the endosperm. These contain little cellulose, and mostly other polysaccharides.

Acid Growth

Acid growth refers to the ability of plant cells and plant cell walls to elongate or expand quickly at low (acidic) pH. This form of growth does not involve an increase in cell number. During acid growth, plant cells enlarge rapidly because the cell walls are made more extensible by expansin, a pH-dependent wall-loosening protein. Expansin loosens the network-like connections between cellulose microfibrils within the cell wall, which allows the cell volume to increase by turgor and osmosis. A typical sequence leading up to this would involve the introduction of a plant hormone (auxin, for example) that causes protons (H^+ ions) to be pumped out of the cell into the cell wall. As a result, the cell wall solution becomes more acidic. This activates expansin activity, causing the wall to become more extensible and to undergo wall stress relaxation, which enables the cell to take up water and to expand.

Cellulose

Cellulose is an organic compound with the formula $(C_6H_{10}O_5)_n$, a polysaccharide consisting of a linear chain of several hundred to many thousands of β(1→4) linkedD-glucose units. Cellulose is an important structural component of the primary cell wall of green plants, many forms of algae and the oomycetes. Some species of bacteria secrete it to form biofilms. Cellulose is the most abundant organic polymer on Earth. The cellulose content of cotton fiber is 90%, that of wood is 40–50%, and that of dried hemp is approximately 57%.

Cellulose is mainly used to produce paperboard and paper. Smaller quantities are converted into a wide variety of derivative products such as cellophane and rayon. Conversion of cellulose from energy crops into biofuels such as cellulosic ethanol is under investigation as an alternative fuel source. Cellulose for industrial use is mainly obtained from wood pulp and cotton.

Some animals, particularly ruminants and termites, can digest cellulose with the help of symbiotic micro-organisms that live in their guts, such as *Trichonympha*. In human nutrition, cellulose is a non-digestible constituent of insolubledietary fiber, acting as a hydrophilicbulking agent for feces and potentially aiding in defecation.

History

Cellulose was discovered in 1838 by the French chemist Anselme Payen, who isolated it from plant matter and determined its chemical formula. Cellulose was used to produce the first successful thermoplastic polymer, celluloid, by Hyatt Manufacturing Company in 1870. Production of rayon ("artificial silk") from cellulose began in the 1890s and cellophane was invented in 1912. Hermann Staudinger determined the polymer structure of cellulose in 1920. The compound was first chemically synthesized (without the use of any biologically derived enzymes) in 1992, by Kobayashi and Shoda.

Structure and Properties

Cellulose has no taste, is odorless, is hydrophilic with the contact angle of 20–30 degrees, is insoluble in water and most organic solvents, is chiral and is biodegradable. It was shown to melt at 467 °C in 2016. It can be broken down chemically into its glucose units by treating it with concentrated mineral acids at high temperature.

Cellulose is derived from D-glucose units, which condense through β(1→4)-glycosidic bonds. This linkage motif contrasts with that for α(1→4)-glycosidic bonds present in starch and glycogen. Cellulose is a straight chain polymer: unlike starch, no coiling or branching occurs, and the molecule adopts an extended and rather stiff rod-like conformation, aided by the equatorial conformation of the glucose residues. The multiple hydroxyl groups on the glucose from one chain form hydrogen bonds with oxygen atoms on the same or on a neighbor chain, holding the chains firmly together side-by-side and forming *microfibrils* with high tensile strength. This confers tensile strength in cell walls, where cellulose microfibrils are meshed into a polysaccharide *matrix*.

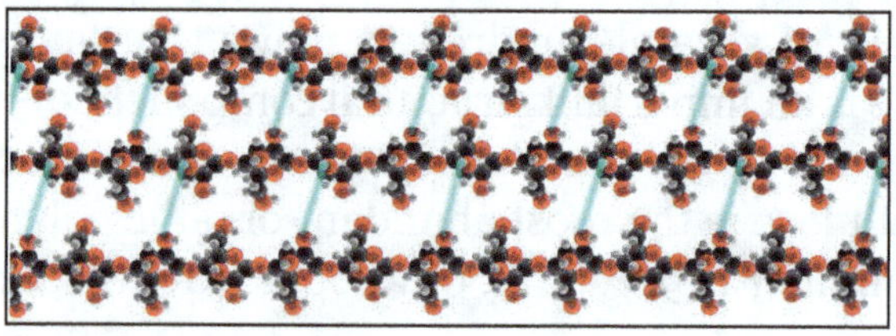

A triple strand of cellulose showing the hydrogen bonds (cyan lines) between glucose strands

Compared to starch, cellulose is also much more crystalline. Whereas starch undergoes a crystalline to amorphous transition when heated beyond 60–70 °C in water (as in cooking), cellulose requires a temperature of 320 °C and pressure of 25 MPa to become amorphous in water.

Cotton fibres represent the purest natural form of cellulose, containing more than 90% of this polysaccharide.

Several different crystalline structures of cellulose are known, corresponding to the location of hydrogen bonds between and within strands. Natural cellulose is cellulose I, with structures I_α and I_β. Cellulose produced by bacteria and algae is enriched in I_α while cellulose of higher plants consists mainly of I_β. Cellulose in regenerated cellulose fibers is cellulose II. The conversion of cellulose I to cellulose II is irreversible, suggesting that cellulose I ismetastable and cellulose II is stable. With various chemical treatments it is possible to produce the structures cellulose III and cellulose IV.

Many properties of cellulose depend on its chain length or degree of polymerization, the number of glucose units that make up one polymer molecule. Cellulose from wood pulp has typical chain lengths between 300 and 1700 units; cotton and other plant fibers as well as bacterial cellulose have chain lengths ranging from 800 to 10,000 units. Molecules with very small chain length resulting from the breakdown of cellulose are known as cellodextrins; in contrast to long-chain cellulose, cellodextrins are typically soluble in water and organic solvents.

Plant-derived cellulose is usually found in a mixture with hemicellulose, lignin, pectin and other substances, while bacterial cellulose is quite pure, has a much higher water content and higher tensile strength due to higher chain lengths.

Cellulose is soluble in Schweizer's reagent, cupriethylenediamine (CED), cadmiumethylenediamine (Cadoxen), *N*-methylmorpholine *N*-oxide, and lithium chloride / dimethylacetamide. This is used in the production of regenerated celluloses (such as viscose and cellophane) from dissolving pulp. Cellulose is also soluble in many kinds of ionic liquids.

Cellulose consists of crystalline and amorphous regions. By treating it with strong acid, the amorphous regions can be broken up, thereby producing nanocrystalline cellulose, a novel material with many desirable properties. Recently, nanocrystalline cellulose was used as the filler phase in bio-based polymer matrices to produce nanocomposites with superior thermal and mechanical properties.

Processing

Assay

Given a cellulose-containing material, the carbohydrate portion that does not dissolve in a 17.5% solution of sodium hydroxide at 20 °C is *α cellulose*, which is true cellulose. Acidification of the extract precipitates *β cellulose*. The portion that dissolves in base but does not precipitate with acid is *γ cellulose*.

Cellulose can be assayed using a method described by Updegraff in 1969, where the fiber is dissolved in acetic and nitric acid to remove lignin, hemicellulose, and xylosans. The resulting cellulose is allowed to react with anthrone in sulfuric acid. The resulting coloured compound is assayed spectrophotometrically at a wavelength of approximately 635 nm.

In addition, cellulose is represented by the difference between acid detergent fiber (ADF) and acid detergent lignin (ADL).

Luminescent conjugated oligothiophenes can also be used to detect cellulose using fluorescence microscopy or spectrofluorometric methods.

Biosynthesis

In vascular plants cellulose is synthesized at the plasma membrane by rosette terminal complexes (RTCs). The RTCs are hexameric protein structures, approximately 25 nm in diameter, that contain the cellulose synthase enzymes that synthesise the individual cellulose chains. Each RTC floats in the cell's plasma membrane and "spins" a microfibril into the cell wall.

RTCs contain at least three different cellulose synthases, encoded by *CesA* genes, in an unknown stoichiometry. Separate sets of *CesA* genes are involved in primary and secondary cell wall biosynthesis. There are known to be about seven subfamilies in the *CesA* superfamily. These cellulose synthases use UDP-glucose to form the β(1→4)-linked cellulose.

Cellulose synthesis requires chain initiation and elongation, and the two processes are separate. *CesA* glucosyltransferase initiates cellulose polymerization using a steroid primer, sitosterol-beta-glucoside, and UDP-glucose.Cellulose synthase utilizes UDP-D-glucose precursors to elongate the growing cellulose chain. A cellulase may function to cleave the primer from the mature chain.

Cellulose is also synthesised by animals, particularly in the tests of ascidians (where the cellulose was historically termed "tunicine") although it is also a minor component of mammalianconnective tissue.

Breakdown (Cellulolysis)

Cellulolysis is the process of breaking down cellulose into smaller polysaccharides called cellodextrins or completely into glucose units; this is a hydrolysis reaction. Because cellulose molecules bind strongly to each other, cellulolysis is relatively difficult compared to the breakdown of other polysaccharides. However, this process can be significantly intensified in a proper solvent, e.g. in an ionic liquid.

Most mammals have limited ability to digest dietary fiber such as cellulose. Some ruminants like cows and sheep contain certain symbioticanaerobic bacteria (like *Cellulomonas*) in the flora of the rumen, and these bacteria produce enzymes called cellulases that help the microorganism to digest cellulose; the breakdown products are then used by the bacteria for proliferation. The bacterial mass is later digested by the ruminant in its digestive system (stomach and small intestine). Horses use cellulose in their diet by fermentation in their hindgut via symbiotic bacteria which produce cellulase to digest cellulose. Similarly, some termites contain in their hindguts certain flagellateprotozoa producing such enzymes, whereas others contain bacteria or may produce cellulase.

The enzymes used to cleave the glycosidic linkage in cellulose are glycoside hydrolases including endo-acting cellulases and exo-acting glucosidases. Such enzymes are usually secreted as part of multienzyme complexes that may include dockerins and carbohydrate-binding modules.

Breakdown (Thermolysis)

At temperatures above 350 °C, cellulose undergoes thermolysis (also called 'pyrolysis'), decomposing into solid char, vapors, aerosols, and gases such as carbon dioxide. Maximum yield of vapors which condense to a liquid called *bio-oil* is obtained at 500 °C.

Semi-crystalline cellulose polymers react at pyrolysis temperatures (350–600 °C) in a few seconds; this transformation has been shown to occur via a solid-to-liquid-to-vapor transition, with the liquid (called *intermediate liquid cellulose* or *molten cellulose*) existing for only a fraction of a second. Glycosidic bond cleavage produces short cellulose chains of two-to-seven monomers comprising the melt. Vapor bubbling of intermediate liquid cellulose produces aerosols, which consist of short chain anhydro-oligomers derived from the melt.

Continuing decomposition of molten cellulose produces volatile compounds including levoglucosan, furans, pyrans, light oxygenates and gases via primary reactions. Within thick cellulose samples, volatile compounds such as levoglucosan undergo 'secondary reactions' to volatile products including pyrans and light oxygenates such as glycolaldehyde.

Hemicellulose

Hemicellulose is a polysaccharide related to cellulose that comprises about 20% of the biomass of most plants. In contrast to cellulose, hemicellulose is derived from several sugars in addition to glucose, especially xylose but also including mannose, galactose, rhamnose, and arabinose. Hemicellulose consists of shorter chains – between 500 and 3000 sugar units. Furthermore, hemicellulose is branched, whereas cellulose is unbranched.

Derivatives

The hydroxyl groups (-OH) of cellulose can be partially or fully reacted with various reagents to afford derivatives with useful properties like mainly cellulose esters and cellulose ethers (-OR). In principle, though not always in current industrial practice, cellulosic polymers are renewable resources.

Ester derivatives include:

Cellulose ester	Reagent	Example	Reagent	Group R
Organic esters	Organic acids	Cellulose acetate	Acetic acid and acetic anhydride	H or $-(C=O)CH_3$
		Cellulose triacetate	Acetic acid and acetic anhydride	$-(C=O)CH_3$
		Cellulose propionate	Propanoic acid	H or $-(C=O)CH_2CH_3$
		Cellulose acetate propionate (CAP)	Acetic acid and propanoic acid	H or $-(C=O)CH_3$ or $-(C=O)CH_2CH_3$
		Cellulose acetate butyrate (CAB)	Acetic acid and butyric acid	H or $-(C=O)CH_3$ or $-(C=O)CH_2CH_2CH_3$
Inorganic esters	Inorganic acids	Nitrocellulose (cellulose nitrate)	Nitric acid or another powerful nitrating agent	H or $-NO_2$
		Cellulose sulfate	Sulfuric acid or another powerful sulfuring agent	H or $-SO_3H$

The cellulose acetate and cellulose triacetate are film- and fiber-forming materials that find a variety of uses. The nitrocellulose was initially used as an explosive and was an early film forming material. With camphor, nitrocellulose gives celluloid.

Ether derivatives include:

Cellulose ethers	Reagent	Example	Reagent	Group R = H or	Water solubility	Application	E number
Alkyl	Halogenoalkanes	Methylcellulose	Chloromethane	$-CH_3$	Cold water-soluble		E461
		Ethylcellulose	Chloroethane	$-CH_2CH_3$	Water-insoluble	A commercial thermoplastic used in coatings, inks, binders, and controlled-release drug tablets	E462
		Ethyl methyl cellulose	Chloromethane and chloroethane	$-CH_3$ or $-CH_2CH_3$			E465
Hydroxyalkyl	Epoxides	Hydroxyethyl cellulose	Ethylene oxide	$-CH_2CH_2OH$	Cold/hot water-soluble	Gelling and thickening agent	
		Hydroxypropyl cellulose (HPC)	Propylene oxide	$-CH_2CH(OH)CH_3$	Cold water-soluble		E463
		Hydroxyethyl methyl cellulose	Chloromethane and ethylene oxide	$-CH_3$ or $-CH_2CH_2OH$	Cold water-soluble	Production of cellulose films	
		Hydroxypropyl methyl cellulose (HPMC)	Chloromethane and propylene oxide	$-CH_3$ or $-CH_2CH(OH)CH_3$	Cold water-soluble	Viscosity modifier, gelling, foaming and binding agent	E464
		Ethyl hydroxyethyl cellulose	Chloroethane and ethylene oxide	$-CH_2CH_3$ or $-CH_2CH_2OH$			E467
Carboxyalkyl	Halogenated carboxylic acids	Carboxymethyl cellulose (CMC)	Chloroacetic acid	$-CH_2COOH$	Cold/Hot water-soluble	Often used as its sodiumsalt, sodium carboxymethyl cellulose (NaCMC)	E466

The sodium carboxymethyl cellulose can be cross-linked to give the croscarmellose sodium (E468) for use as a disintegrant in pharmaceutical formulations.

Applications

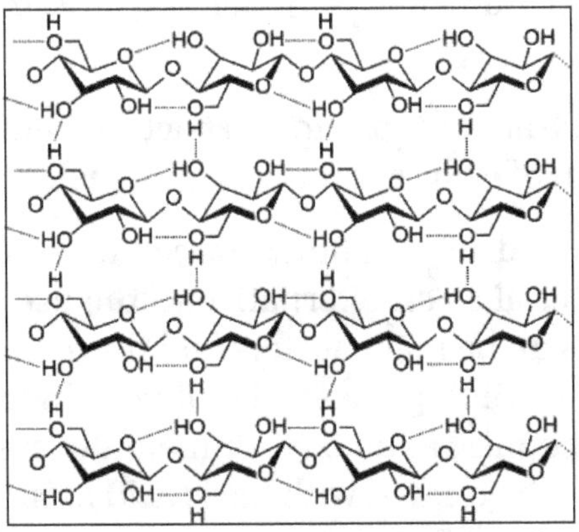

A strand of cellulose (conformation I_α), showing the hydrogen bonds (dashed) within and between cellulose molecules.

Cellulose for industrial use is mainly obtained from wood pulp and cotton. The kraft process is used to separate cellulose from lignin, another major component of plant matter:

- Paper products: Cellulose is the major constituent of paper, paperboard, and card stock.
- Fibers: Cellulose is the main ingredient of textiles made from cotton, linen, and other plant fibers. It can be turned into rayon, an important fiber that has been used for textiles since the beginning of the 20th century. Both cellophane and rayon are known as "regenerated cellulose fibers"; they are identical to cellulose in chemical structure and are usually made from dissolving pulp via viscose. A more recent and environmentally friendly method to produce a form of rayon is the Lyocell process.
- Consumables: Microcrystalline cellulose (E460i) and powdered cellulose (E460ii) are used as inactive fillers in drug tablets and a wide range of soluble cellulose derivatives, E numbers E461 to E469, are used as emulsifiers, thickeners and stabilizers in processed foods. Cellulose powder is, for example, used in Parmesan cheese to prevent caking inside the package. Cellulose occurs naturally in some foods and is an additive in manufactured foods, contributing an indigestible component used for texture and bulk, potentially aiding in defecation.
- Science: Cellulose is used in the laboratory as a stationary phase for thin layer chromatography. Cellulose fibers are also used in liquid filtration, sometimes in combination with diatomaceous earth or other filtration media, to create a filter bed of inert material.
- Energy crops.

The major combustible component of non-food energy crops is cellulose, with lignin second. Non-food energy crops produce more usable energy than edible energy crops (which have a large starch component), but still compete with food crops for agricultural land and water resources. Typical non-food energy crops include industrial hemp (though outlawed in some countries), switchgrass, *Miscanthus*, *Salix* (willow), and *Populus* (poplar) species:

- Biofuel: TU-103, a strain of *Clostridium* bacteria found in zebra waste, can convert nearly any form of cellulose into butanol fuel.
- Building material: Hydroxyl bonding of cellulose in water produces a sprayable, moldable material as an alternative to the use of plastics and resins. The recyclable material can be made water- and fire-resistant. It provides sufficient strength for use as a building material.Cellulose insulation made from recycled paper is becoming popular as an environmentally preferable material for building insulation. It can be treated with boric acid as a fire retardant.
- Miscellaneous: Cellulose can be converted into cellophane, a thin transparent film. It is the base material for the celluloid that was used for photographic and movie films until the mid-1930s. Cellulose is used to make water-soluble adhesives and binders such as methyl cellulose and carboxymethyl cellulose which are used in wallpaper paste. Cellulose is further used to make hydrophilic and highly absorbent sponges. Cellulose is the raw material in the manufacture of nitrocellulose (cellulose nitrate) which is used in smokeless gunpowder.

Hemicellulose

A hemicellulose (also known as polyose) is any of several heteropolymers (matrix polysaccharides), such as arabinoxylans, present along with cellulose in almost all plant cell walls. While cellulose is crystalline, strong, and resistant to hydrolysis, hemicellulose has a random, amorphous structure with little strength. It is easily hydrolyzed by dilute acid or base as well as myriad hemicellulase enzymes.

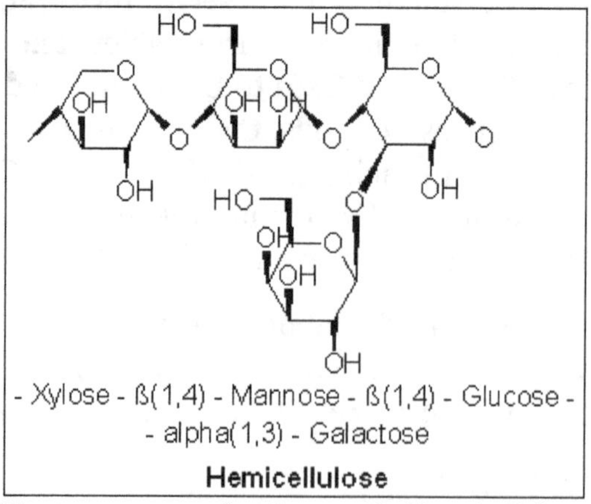

Most common molecular motif of hemicellulose.

Composition

Hemicelluloses include xylan, glucuronoxylan, arabinoxylan, glucomannan, and xyloglucan.

These polysaccharides contain many different sugar monomers. In contrast, cellulose contains only anhydrous glucose. For instance, besides glucose, sugar monomers in hemicellulose can include xylose, mannose, galactose, rhamnose, and arabinose. Hemicelluloses contain most of the D-pentose sugars, and occasionally small amounts of L-sugars as well. Xylose is in most cases the sugar monomer present in the largest amount, although in softwoods mannose can be the most abundant sugar. Not only regular sugars can be found in hemicellulose, but also their acidified form, for instance glucuronic acid and galacturonic acid can be present. Hemicellulose is often associated with cellulose, but it has different composition.

Structural Comparison to Cellulose

Unlike cellulose, hemicellulose (also a polysaccharide) consists of shorter chains – 500–3,000 sugar units as opposed to 7,000–15,000 glucose molecules per polymer, as seen in cellulose. In addition, hemicellulose is a branched polymer, while cellulose is unbranched.

Native Structure

Hemicelluloses are embedded in the cell walls of plants, sometimes in chains that form a 'ground' - they bind with pectin to cellulose to form a network of cross-linked fibres.

Biosynthesis

Hemicelluloses are synthesised from sugar nucleotides in the cell's Golgi apparatus. Two models explain their synthesis: 1) a '2 component model' where modification occurs at two transmembrane proteins, and 2) a '1 component model' where modification occurs only at one transmembrane protein. After synthesis, hemicelluloses are transported to the plasma membrane via Golgi vesicles.

Applications

As percent content of hemicellulose increases in animal feed, the voluntary feed intake decreases.

Hemicellulose is represented by the difference between neutral detergent fiber (NDF) and acid detergent fiber (ADF).

Functions

Microfibrils are cross-linked together by hemicellulose homopolymers. Lignins assist and strengthen the attachment of hemicelluloses to microfibrils.

Hemicellulose from Trees

Hemicellulose found in hardwood trees is predominantly xylan with some glucomannan, while in softwoods it is mainly rich in galactoglucomannan and contains only a small amount of xylan. The average molecular weight is lower than that of cellulose at less than 30,000, as opposed to the 100,000 average molecular weight reported for cellulose.

Pectin

Pectin, a bio polymer of (among other constituents) D-galacturonic acid, shown here in a powder form.

Pectin is a structural heteropolysaccharide contained in the primary cell walls of terrestrial plants. It was first isolated and described in 1825 by Henri Braconnot. It is produced commercially as a white to light brown powder, mainly extracted from citrus fruits, and is used in food as a gelling agent, particularly in jams and jellies. It is also used in dessert fillings, medicines, sweets, as a stabilizer in fruit juices and milk drinks, and as a source of dietary fiber.

Biology

In plant biology, pectin consists of a complex set of polysaccharides that are present in most primary cell walls and are particularly abundant in the non-woody parts of terrestrial plants. Pectin is a major component of the middle lamella, where it helps to bind cells together, but is also found in primary cell walls.

The amount, structure and chemical composition of pectin differs among plants, within a plant over time, and in various parts of a plant. Pectin is an important cell wall polysaccharide that allows primary cell wall extension and plant growth. During fruit ripening, pectin is broken down by the enzymespectinase and pectinesterase, in which process the fruit becomes softer as the middle lamellae break down and cells become separated from each other. A similar process of cell separation caused by the breakdown of pectin occurs in the abscission zone of the petioles of deciduous plants at leaf fall.

Pectin is a natural part of the human diet, but does not contribute significantly to nutrition. The daily intake of pectin from fruits and vegetables can be estimated to be around 5 g (assuming consumption of approximately 500 g fruits and vegetables per day).

In human digestion, pectin binds to cholesterol in the gastrointestinal tract and slows glucose absorption by trapping carbohydrates. Pectin is thus a soluble dietary fiber.

Pectin has been observed to have some function in DNA repair of plants. Pectinaceous surface pellicles, which are rich in pectin, create a mucilage layer that holds in dew that helps the cell repair its DNA.

Consumption of pectin has been shown to reduce blood cholesterol levels. The mechanism appears to be an increase of viscosity in the intestinal tract, leading to a reduced absorption of cholesterol from bile or food. In the large intestine and colon, microorganisms degrade pectin and liberate short-chain fatty acids that have positive influence on health (prebiotic effect).

Human Metabolites

A study found a mean of 4.5 ppmmethanol in the exhaled breath of subjects. The mean endogenous methanol production in humans of 0.45 g/d may be metabolized from pectin found in fruit; one kilogram of apple produces up to 1.4 g methanol.Methanol is poisonous to the central nervous system and may cause blindness, coma, and death. However, in small amounts, methanol is a natural endogenous compound found in normal, healthy human individuals.

Chemistry

Pectins, also known as pectic polysaccharides, are rich in galacturonic acid. Several distinct polysaccharides have been identified and characterised within the pectic group. Homogalacturonans are linear chains of α-(1–4)-linked D-galacturonic acid. Substituted galacturonans are characterized by the presence of saccharide appendant residues (such as D-xylose or D-apiose in the respective cases of xylogalacturonan and apiogalacturonan) branching from a backbone of D-galacturonic acid residues. Rhamnogalacturonan I pectins (RG-I) contain a backbone of the repeating disaccharide: 4)-α-D-galacturonic acid-(1,2)-α-L-rhamnose-(1. From many of the rhamnose residues, sidechains of various neutral sugars branch off. The neutral sugars are mainly D-galactose, L-arabinose and D-xylose, with the types and proportions of neutral sugars varying with the origin of pectin.

Another structural type of pectin is rhamnogalacturonan II (RG-II), which is a less frequent, complex, highly branched polysaccharide.Rhamnogalacturonan II is classified by some authors within the group of substituted galacturonans since the rhamnogalacturonan II backbone is made exclusively of D-galacturonic acid units.

Isolated pectin has a molecular weight of typically 60,000–130,000 g/mol, varying with origin and extraction conditions.

In nature, around 80 percent of carboxyl groups of galacturonic acid are esterified with methanol. This proportion is decreased to a varying degree during pectin extraction. The ratio of esterified to non-esterified galacturonic acid determines the behavior of pectin in food applications. This is why pectins are classified as high- vs. low-ester pectins (short HM vs. LM-pectins), with more or less than half of all the galacturonic acid esterified.

The non-esterified galacturonic acid units can be either free acids (carboxyl groups) or salts with sodium, potassium, or calcium. The salts of partially esterified pectins are called pectinates, if the degree of esterification is below 5 percent the salts are called pectates, the insoluble acid form, pectic acid.

Some plants such as sugar beet, potatoes and pears contain pectins with acetylated galacturonic acid in addition to methyl esters. Acetylation prevents gel-formation but increases the stabilising and emulsifying effects of pectin.

Amidated pectin is a modified form of pectin. Here, some of the galacturonic acid is converted with ammonia to carboxylic acidamide. These pectins are more tolerant of varying calcium concentrations that occur in use.

To prepare a pectin-gel, the ingredients are heated, dissolving the pectin. Upon cooling below gelling temperature, a gel starts to form. If gel formation is too strong, syneresis or a granular texture are the result, whilst weak gelling leads to excessively soft gels. Pectins gel according to specific parameters, such as sugar, pH and bivalent salts (especially Ca^{2+}).

In high-ester pectins at soluble solids content above 60% and a pH-value between 2.8 and 3.6, hydrogen bonds and hydrophobic interactions bind the individual pectin chains together. These bonds form as water is bound by sugar and forces pectin strands to stick together. These form a 3-dimensional molecular net that creates the macromolecular gel. The gelling-mechanism is called a low-water-activity gel or sugar-acid-pectin gel.

In low-ester pectins, ionic bridges are formed between calcium ions and the ionised carboxyl groups of the galacturonic acid. This is idealised in the so-called "egg box-model". Low-ester pectins need calcium to form a gel, but can do so at lower soluble solids and higher pH-values than high-ester pectins. Normally low-ester pectins form gels with a range of pH from 2.6 to 7.0 and with a soluble solids content between 10 and 70%.

Amidated pectins behave like low-ester pectins but need less calcium and are more tolerant of excess calcium. Also, gels from amidated pectin are thermo-reversible; they can be heated and after cooling solidify again, whereas conventional pectin-gels will afterwards remain liquid.

High-ester pectins set at higher temperatures than low-ester pectins. However, gelling reactions with calcium increase as the degree of esterification falls. Similarly, lower pH-values or higher soluble solids (normally sugars) increase gelling speed. Suitable pectins can therefore be selected for jams and for jellies, or for higher sugar confectionery jellies.

Sources and Production

Pears, apples, guavas, quince, plums, gooseberries, and oranges and other citrus fruits contain large amounts of pectin, while soft fruits like cherries, grapes, and strawberries contain small amounts of pectin.

Typical levels of pectin in plants are (fresh weight):

- Apples, 1–1.5%
- Apricots, 1%
- Cherries, 0.4%
- Oranges, 0.5–3.5%
- Carrots approx. 1.4%
- Citrus peels, 30%

The main raw materials for pectin production are dried citrus peel or apple pomace, both by-products of juice production. Pomace from sugar beet is also used to a small extent.

From these materials, pectin is extracted by adding hot dilute acid at pH-values from 1.5 – 3.5. During several hours of extraction, the protopectin loses some of its branching and chain length and goes into solution. After filtering, the extract is concentrated in vacuum and the pectin then precipitated by adding ethanol or isopropanol. An old technique of precipitating pectin with aluminium salts is no longer used (apart from alcohols and polyvalent cations, pectin also precipitates with proteins and detergents).

Alcohol-precipitated pectin is then separated, washed and dried. Treating the initial pectin with dilute acid leads to low-esterified pectins. When this process includes ammonium hydroxide, amidated pectins are obtained. After drying and milling, pectin is usually standardised with sugar and sometimes calcium salts or organic acids to have optimum performance in a particular application.

Uses

The main use for pectin (vegetable agglutinate) is as a gelling agent, thickening agent and stabilizer in food. The classical application is giving the jelly-like consistency to jams or marmalades, which would otherwise be sweet juices. Pectin also reduces syneresis in jams and marmalades and increases the gel strength of low calorie jams. For household use, pectin is an ingredient in gelling sugar (also known as "jam sugar") where it is diluted to the right concentration with sugar and some citric acid to adjust pH. In some countries, pectin is also available as a solution or an extract, or as a blended powder, for home jam making. For conventional jams and marmalades that contain above 60% sugar and soluble fruit solids, high-ester pectins are used. With low-ester pectins and amidated pectins less sugar is needed, so that diet products can be made.

Pectin is used in confectionery jellies to give a good gel structure, a clean bite and to confer a good flavour release. Pectin can also be used to stabilize acidic protein drinks, such as drinking yogurt, to improve the mouth-feel and the pulp stability in juice based drinks and as a fat substitute in baked goods. Typical levels of pectin used as a food additive are between 0.5 and 1.0% – this is about the same amount of pectin as in fresh fruit.

In medicine, pectin increases viscosity and volume of stool so that it is used against constipation and diarrhea. Until 2002, it was one of the main ingredients used in Kaopectate a medication to combat diarrhea, along with kaolinite. It has been used in gentle heavy metal removal from biological systems. Pectin is also used in throat lozenges as a demulcent.

In cosmetic products, pectin acts as stabilizer. Pectin is also used in wound healing preparations and specialty medical adhesives, such as colostomy devices.

Sriamornsak revealed that pectin could be used in various oral drug delivery platforms, e.g., controlled release systems, gastro-retentive systems, colon-specific delivery systems and mucoadhesive delivery systems, according to its intoxicity and low cost. It was found that pectin from different sources provides different gelling abilities, due to variations in molecular size and chemical composition. Like other natural polymers, a major problem with pectin is inconsistency in reproducibility between samples, which may result in poor reproducibility in drug delivery characteristics.

In ruminant nutrition, depending on the extent of lignification of the cell wall, pectin is up to 90% digestible by bacterial enzymes. Ruminant nutritionists recommend that the digestibility and energy concentration in forages can be improved by increasing pectin concentration in the forage.

In the cigar industry, pectin is considered an excellent substitute for vegetable glue and many cigar smokers and collectors will use pectin for repairing damaged tobacco wrapper leaves on their cigars.

Yablokov *et al.*, writing in Chernobyl: Consequences of the Catastrophe for People and the Environment, quote research conducted by the Ukrainian Center of Radiation Medicine and the Belarusian Institute of Radiation Medicine and Endocrinology, concluded about pectin's radioprotective effects that "adding pectin preparations to the food of inhabitants of the Chernobyl-contaminated regions promotes an effective excretion of incorporated radionuclides" such as cesium-137. The authors report on the positive results of using pectin food additive preparations in a number of clinical studies conducted on children in severely polluted areas, with up to 50% improvement over control groups.

During the Second World War, Allied pilots were provided with maps printed on silk, for navigation in escape and evasion efforts. The printing process at first proved nearly impossible because the several layers of the ink immediately ran, blurring outlines and rendering place names illegible... until the inventor of the maps, Clayton Hutton, mixed a little pectin with the ink and at once the pectin coagulated the ink and prevented it from running. Now, even the smallest topographic features were sharply defined on the map.

Hazardous Metabolites

Fruit consumption is a source of endogenous methanol metabolized from pectin. One kilogram of apples produces up to 1.4 grams of methanol.

Legal Status

At the Joint FAO/WHO Expert Committee Report on Food Additives and in the European Union, no numerical acceptable daily intake (ADI) has been set, as pectin is considered safe.

In the United States, pectin is GRAS – generally recognized as safe. In most foods it can be used according to good manufacturing practices in the levels needed for its application ("quantum satis").

In the International Numbering System (INS), pectin has the number 440. In Europe, pectins are differentiated into the E numbersE440(i) for non-amidated pectins and E440(ii) for amidated pectins. There are specifications in all national and international legislation defining its quality and regulating its use.

History

Pectin was first isolated and described in 1825 by Henri Braconnot, though the action of pectin to make jams and marmalades was known long before. To obtain well set jams from fruits that had little or only poor quality pectin, pectin-rich fruits or their extracts were mixed into the recipe.

During the Industrial Revolution, the makers of fruit preserves turned to producers of apple juice to obtain dried apple pomace that was cooked to extract pectin.

Later, in the 1920s and 1930s, factories were built that commercially extracted pectin from dried apple pomace and later citrus-peel in regions that produced apple juice in both the USA and in Europe.

Pectin was first sold as a liquid extract, but is now most often used as dried powder, which is easier than a liquid to store and handle.

Middle Lamella

The middle lamella is a pectin layer which cements the cell walls of two adjoining plant cells together. It is the first formed layer which is deposited at the time of cytokinesis. The cell plate that is formed during cell division itself develops into middle lamella or lamellum. The middle lamella is made up of calcium and magnesium pectates. In a mature plant cell it is outermost layer of cell wall.

In plants, the pectins form a unified and continuous layer between adjacent cells. Frequently, it is difficult to distinguish the middle lamella from the primary wall, especially in cells that develop thick secondary walls. In such cases, the two adjacent primary walls and the middle lamella, and perhaps the first layer of the secondary wall of each cell, may be called a compound middle lamella. When the middle lamella is degraded by enzymes, as happens during fruit ripening, the adjacent cells will separate.

Lignin

Lignin is a class of complex organic polymers that form important structural materials in the support tissues of vascular plants and some algae. Lignins are particularly important in the formation of cell walls, especially in wood and bark, because they lend rigidity and do not rot easily. Chemically, lignins are cross-linked phenolic polymers.

History

Lignin was first mentioned in 1813 by the Swiss botanist A. P. de Candolle, who described it as a fibrous, tasteless material, insoluble in water and alcohol but soluble in weak alkaline solutions, and which can be precipitated from solution using acid. He named the substance "lignine", which is derived from the Latin word *lignum*, meaning wood. It is one of the most abundant organic polymers on Earth, exceeded only by cellulose. Lignin constitutes 30% of non-fossil organic carbon and 20-35% of the dry mass of wood. The CarboniferousPeriod (geology) is in part defined by the evolution of lignin.

Composition

The composition of lignin varies from species to species. An example of composition from an aspen sample is 63.4% carbon, 5.9% hydrogen, 0.7% ash (mineral components), and 30% oxygen (by difference), corresponding approximately to the formula $(C_{31}H_{34}O_{11})_n$. As a biopolymer, lignin is unusual because of its heterogeneity and lack of a defined primary structure. Its most commonly noted function is the support through strengthening of wood (mainly composed of xylem cells and lignified sclerenchyma fibres) in vascular plants.

Global commercial production of lignin is around 1.1 million metric tons per year and is used in a wide range of low volume, niche applications where the form but not the quality is important.

Biological Function

Lignin fills the spaces in the cell wall between cellulose, hemicellulose, and pectin components, especially in vascular and support tissues: xylemtracheids, vessel elements and sclereid cells. It is covalently linked to hemicellulose and therefore cross-links different plant polysaccharides, conferring mechanical strength to the cell wall and by extension the plant as a whole. It is particularly abundant in compression wood but scarce in tension wood, which are types of reaction wood.

Lignin plays a crucial part in conducting water in plant stems. The polysaccharide components of plant cell walls are highly hydrophilic and thus permeable to water, whereas lignin is more hydrophobic. The crosslinking of polysaccharides by lignin is an obstacle for water absorption to the cell wall. Thus, lignin makes it possible for the plant's vascular tissue to conduct water efficiently. Lignin is present in all vascular plants, but not in bryophytes, supporting the idea that the original function of lignin was restricted to water transport. However, it is present in red algae, which seems to suggest that the common ancestor of plants and red algae also synthesised lignin. This would suggest that its original function was structural; it plays this role in the red alga *Calliarthron*, where it supports joints between calcified segments. Another possibility is that the lignins in red algae and in plants are result of convergent evolution and not of a common origin.

Ecological Function

Lignin plays a significant role in the carbon cycle, sequestering atmospheric carbon into the living tissues of woody perennial vegetation. Lignin is one of the most slowly decomposing components of dead vegetation, contributing a major fraction of the material that becomes humus. The resulting soil humus, in general, holds nutrients onto its surface, and hence increases its cation exchange capacity and moisture retention, hence it increases the productivity of soil.

Economic Significance

Mechanical, or high-yield pulp used to make newsprint contains most of the lignin originally present in the wood. This lignin is responsible for newsprint's yellowing with age. Lignin must be removed from the pulp before high-quality bleached paper can be manufactured.

In sulfite pulping, lignin is removed from wood pulp as sulfonates. These lignosulfonates have several uses:

- Dispersants in high performance cement applications, water treatment formulations and textile dyes.
- Additives in specialty oil field applications and agricultural chemicals.
- Raw materials for several chemicals, such as vanillin, DMSO, ethanol, xylitol sugar, and humic acid.
- Environmentally sustainable dust suppression agent for roads.

Lignin removed via the kraft process is usually burned for its fuel value as part of a concentrated black liquor stream, providing energy to run the mill and its associated processes. Three commercial processes exist to remove lignin from black liquor for higher value uses: LignoBoost (Sweden), LignoForce (Canada), and SLRP (US). Higher quality lignin presents the potential to become the main renewable aromatic resource for the chemical industry in the future, with an addressable market of more than $130bn.

Given that lignin is the most prevalent biopolymer after cellulose and is ubiquitous in the Earth's biosphere, the same economic principles that drive the desire for cellulosic ethanol as a biofuel also call for the investigation of lignin as a feedstock for biofuel production. Lignin can already be burned in furnaces, but interest in the idea of instead chemically converting it to liquid fuel is strong.

Timeline

- 1927: The first investigations into commercial use of lignin were reported by Marathon Corporation, a paper company based in Rothschild, Wisconsin. The first class of products that showed promise were leather tanning agents. The lignin chemical business of Marathon was operated for many years as Marathon Chemicals. It is now known as LignoTech USA, Inc., and is owned by the Norwegian company Borregaard.
- 1998: a German company, Tecnaro, developed a process for turning lignin into a substance, called Arboform, which behaves identically to plastic for injection molding. Therefore, it can be used in place of plastic for several applications. When the item is discarded, it can be burned just like wood.
- 2007: lignin extracted from shrubby willow was successfully used to produce expanded polyurethane foam.
- 2012: it was shown carbon fiber can be produced from lignin instead of from fossil oil.
- 2013: the Flemish Institute for Biotechnology was supervising a trial of 448

poplar trees genetically engineered to produce less lignin so that they would be more suitable for conversion into bio-fuels.

- 2013: researchers from the University of California, Berkeley, demonstrated that lignin derived from a Miscanthus grass could be catalytically degraded into monophenolic products, which could potentially serve as an aromatic chemical feedstock.

Structure

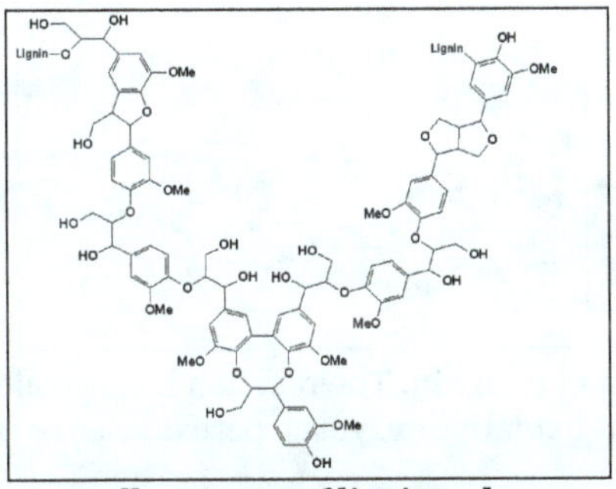

A small segment of lignin polymer

Lignin is a cross-linked racemicmacromolecule with molecular masses in excess of 10,000 u. It is relatively hydrophobic and aromatic in nature. The degree of polymerisation in nature is difficult to measure, since it is fragmented during extraction and the molecule consists of various types of substructures that appear to repeat in a haphazard manner. Different types of lignin have been described depending on the means of isolation.

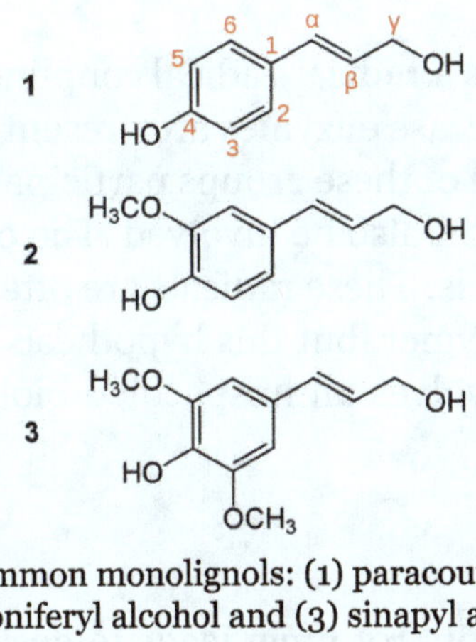

The three common monolignols: (1) paracoumaryl alcohol, (2) coniferyl alcohol and (3) sinapyl alcohol.

There are three monolignolmonomers, methoxylated to various degrees: *p*-coumaryl alcohol, coniferyl alcohol, and sinapyl alcohol. These lignols are incorporated into lignin in the form of the phenylpropanoids*p*-hydroxyphenyl (H), guaiacyl (G), and syringyl (S), respectively.Gymnosperms have a lignin that consists almost entirely of G with small quantities of H. That of dicotyledonousangiosperms is more often than not a mixture of G and S (with very little H), and monocotyledonous lignin is a mixture of all three. Many grasses have mostly G, while some palms have mainly S. All lignins contain small amounts of incomplete or modified monolignols, and other monomers are prominent in non-woody plants.

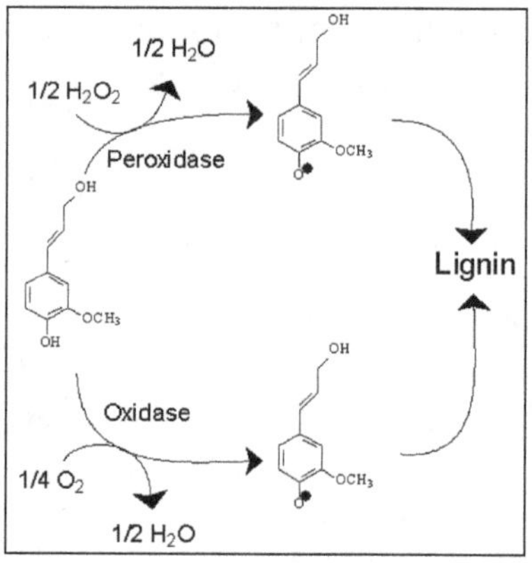

Polymerisation of coniferyl alcohol to lignin. The reaction has two alternative routes catalysed by two different oxidative enzymes, peroxidases or oxidases.

Biosynthesis

Lignin biosynthesis begins in the cytosol with the synthesis of glycosylated monolignols from the amino acidphenylalanine. These first reactions are shared with the phenylpropanoid pathway. The attached glucose renders them water-soluble and less toxic. Once transported through the cell membrane to the apoplast, the glucose is removed and the polymerisation commences. Much about its anabolism is not understood even after more than a century of study.

The polymerisation step, that is a radical-radical coupling, is catalysed by oxidative enzymes. Both peroxidase and laccase enzymes are present in the plantcell walls, and it is not known whether one or both of these groups participates in the polymerisation. Low molecular weight oxidants might also be involved. The oxidative enzyme catalyses the formation of monolignol radicals. These radicals are often said to undergo uncatalyzed coupling to form the lignin polymer, but this hypothesis has been recently challenged. The alternative theory that involves an unspecified biological control is however not widely accepted.

Biodegradation

Biodegradation of lignin by white rot fungi leads to destruction of wood on the forest floor and human-made structures such as fences and wooden buildings. However biodegradation of lignin is a necessary prerequisite for processing biofuel from plant raw materials. Current processing setups show some problematic residuals after processing the digestible or degradable contents. The improving of lignin degradation would drive the output from biofuel processing to better gain or better efficiency.

Lignin is indigestible by animals, which lack the enzymes that can degrade this complex polymer. Some fungi (such as the Dryad's saddle) and bacteria do however biodegrade lignin using so-called ligninases (also named *lignases*). The mechanism of the biodegradation is speculated to involve free radical pathways. Well understood ligninolytic enzymes are manganese peroxidase and lignin peroxidase. Because it is cross-linked with the other cell wall components and has a high molecular weight, lignin minimizes the accessibility of cellulose and hemicellulose to microbial enzymes such as cellobiose dehydrogenase. Hence, in general lignin is associated with reduced digestibility of the overall plant biomass, which helps defend against pathogens and pests. Syringyl (S) lignol is more susceptible to degradation by fungal decay as it has fewer aryl-aryl bonds and a lower redox potential than guaiacyl units. This means that organic matter that is enriched with G lignol (like the bark of woody vascular plants) is more resistant to microbial attack.

Lignin is degraded by micro-organisms including fungi and bacteria. Lignin peroxidase (also "ligninase", EC number 1.14.99) is a hemoprotein firstly isolated from the white-rot fungus *Phanerochaete chrysosporium* with a variety of lignin-degrading reactions, all utilizing hydrogen peroxide as an oxygen source. Other microbial enzymes may be involved in lignin biodegradation, such as manganese peroxidase and the copper-based laccase.

Pyrolysis

Pyrolysis of lignin during the combustion of wood or charcoal production yields a range of products, of which the most characteristic ones are methoxy-substituted phenols. Of those, the most important are guaiacol and syringol and their derivatives; their presence can be used to trace a smoke source to a wood fire. In cooking, lignin in the form of hardwood is an important source of these two chemicals, which impart the characteristic aroma and taste to smoked foods such as barbecue. The main flavor compounds of smoked ham are guaiacol, and its 4-, 5-, and 6-methyl derivatives as well as 2,6-dimethylphenol. These compounds are produced by thermal breakdown of lignin in the wood used in the smokehouse.

Chemical Analysis

The conventional method for lignin quantitation in the pulp industry is the Klason lignin and acid-soluble lignin test, which is standardized according to SCAN or NREL procedure. The cellulose is first decrystallized and partially depolymerized into oligomers by keeping the sample in 72% sulfuric acid at 30 C for 1 h. Then, the acid is diluted to 4% by adding water, and the depolymerization is completed by either boiling (100 °C) for 4 h or pressure cooking at 2 bar (124 °C) for 1 h. The acid is washed out and the sample dried. The residue that remains is termed Klason lignin. A part of the lignin, acid-soluble lignin (ASL) dissolves in the acid. ASL is quantified by the intensity of its UV absorption peak at 280 nm. The method is suited for wood lignins, but not equally well for varied lignins from different sources. The carbohydrate composition may be also analyzed from the Klason liquors, although there may be sugar breakdown products (furfural and 5-hydroxymethylfurfural).

A solution of hydrochloric acid and phloroglucinol is used for the detection of lignin (Wiesner test). A brilliant red color develops, owing to the presence of coniferaldehyde groups in the lignin.

Thioglycolysis is an analytical technique for lignin quantitation. Lignin structure can also be studied by computational simulation.

Thermochemolysis (chemical break down of a substance under vacuum and at high temperature) with tetramethylammonium hydroxide (TMAH) has also been used to analyse the ratios of lignols with fungal decay as well the ratio of the carboxylic acid (Ad) to aldehyde (Al) forms of the lignols (Ad/Al). Increases in the (Ad/Al) value indicate an oxidative cleavage reaction has occurred on the alkyl lignin side chain which has been shown to be a step in the decay of wood by many white-rot and some soft rot fungi.

Solid state ^{13}C NMR has been used to look at the concentrations of lignin, as well as other major components in wood e.g. cellulose, and how that changes with microbial decay. Conventional solution-state NMR for lignin is possible. However, many intact lignins have a crosslinked, very high molar-mass fraction that is difficult to dissolve even for functionalization.

Plasmodesmata

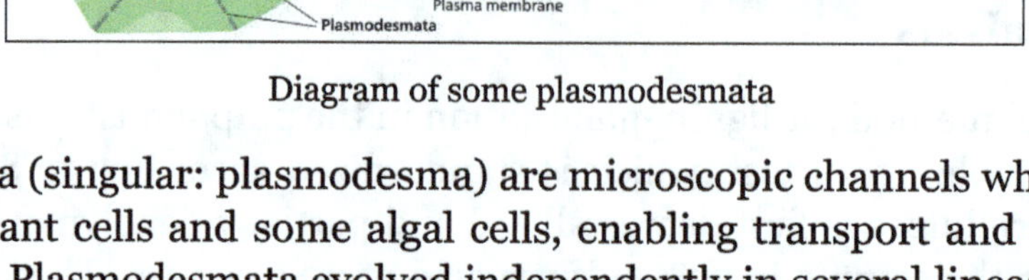

Diagram of some plasmodesmata

Plasmodesmata (singular: plasmodesma) are microscopic channels which traverse the cell walls of plant cells and some algal cells, enabling transport and communication between them. Plasmodesmata evolved independently in several lineages, and species that have these structures include members of the Charophyceae, Charales, Coleochaetales and Phaeophyceae (which are all algae), as well as all embryophytes, better known as land plants. Unlike animal cells, almost every plant cell is surrounded by a polysaccharidecell wall. Neighbouring plant cells are therefore separated by a pair of cell walls and the intervening middle lamella, forming an extracellular domain known as the apoplast. Although cell walls are permeable to small soluble proteins and other solutes, plasmodesmata enable direct, regulated, symplastic transport of substances between cells. There are two forms of plasmodesmata: primary plasmodesmata, which are formed during cell division, and secondary plasmodesmata, which can form between mature cells.

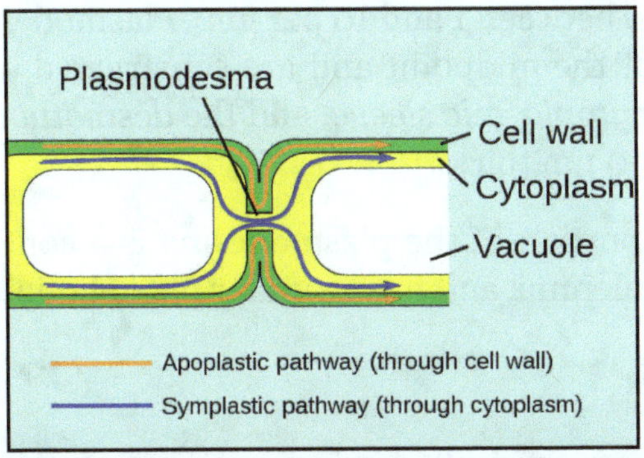

Plasmodesmata allow molecules to travel between plant cells through the symplastic pathway

Similar structures, called gap junctions and membrane nanotubes, interconnect animal cells and stromules form between plastids in plant cells.

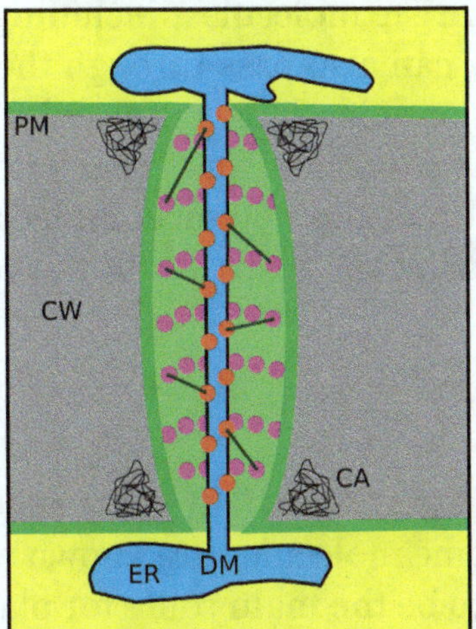

The structure of a primary plasmodesma. CW=Cell wall CA=Callose PM=Plasma membrane ER=Endoplasmic reticulum DM=Desmotubule Red circles=Actin Purple circles and spokes=Other unidentified proteins.

Formation

Primary plasmodesmata are formed when portions of the endoplasmic reticulum are trapped across the middle lamella as new cell wall is laid down between two newly divided plant cells and these eventually become the cytoplasmic connections between cells. Here the wall is not thickened further, and depressions or thin areas known as pits are formed in the walls. Pits normally pair up between adjacent cells. Plasmodesmata can also be inserted into existing cell walls between non-dividing cells (secondary plasmodesmata).

Structure

Plasmodesmatal Plasma Membrane

A typical plant cell may have between 10^3 and 10^5 plasmodesmata connecting it with adjacent cells equating to between 1 and 10 per μm^2. Plasmodesmata are approximately 50-60 nm in diameter at the midpoint and are constructed of three main layers, the plasma membrane, the *cytoplasmic sleeve*, and the *desmotubule*. They can transverse cell walls that are up to 90 nm thick.

The plasma membrane portion of the plasmodesma is a continuous extension of the cell membrane or plasmalemma and has a similar phospholipid bilayer structure.

Cytoplasmic Sleeve

The cytoplasmic sleeve is a fluid-filled space enclosed by the plasmalemma and is a continuous extension of the cytosol. Trafficking of molecules and ions through plasmodesmata occurs through this space. Smaller molecules (e.g. sugars and amino acids) and ions can easily pass through plasmodesmata by diffusion without the need for additional chemical energy. Larger molecules, including proteins (for example green fluorescent protein) and RNA, can also pass through the cytoplasmic sleeve diffusively. Plasmodesmatal transport of some larger molecules is facilitated by mechanisms that are currently unknown. One mechanism of regulation of the permeability of plasmodesmata is the accumulation of the polysaccharidecallose around the neck region to form a collar, thereby reducing the diameter of the pore available for transport of substances.

Desmotubule

The desmotubule is a tube of appressed (flattened) endoplasmic reticulum that runs between two adjacent cells. Some molecules are known to be transported through this channel, but it is not thought to be the main route for plasmodesmatal transport.

Around the desmotubule and the plasma membrane areas of an electron dense material have been seen, often joined together by spoke-like structures that seem to split the plasmodesma into smaller channels. These structures may be composed of myosin and actin, which are part of the cell's cytoskeleton. If this is the case these proteins could be used in the selective transport of large molecules between the two cells.

Transport

Plasmodesmata have been shown to transport proteins (including transcription factors), short interfering RNA, messenger RNA, viroids, and viral genomes from cell to cell. One example of a viral movement proteins is the tobacco mosaic virus MP-30. MP-30 is thought to bind to the virus's own genome and shuttle it from infected cells to uninfected cells through plasmodesmata.Flowering Locus T protein moves from leaves to the shoot apical meristem through plasmodesmata to initiate flowering.

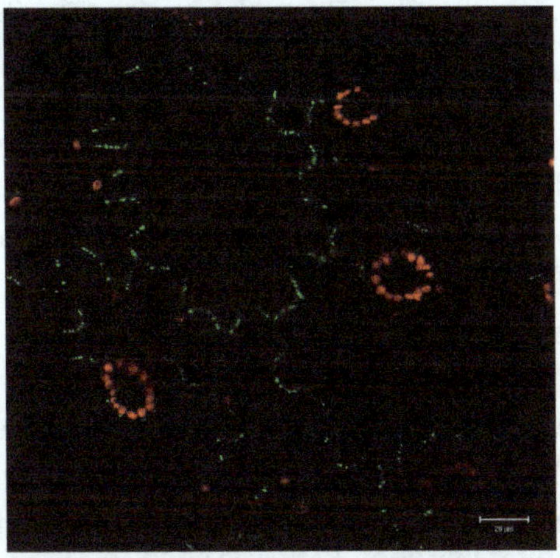

Tobacco mosaic virus movement protein 30 localizes to plasmodesmata.

Plasmodesmata are also used by cells in phloem, and symplastic transport is used to regulate the sieve-tube cells by the companion cells.

The size of molecules that can pass through plasmodesmata is determined by the size exclusion limit. This limit is highly variable and is subject to active modification. For example, MP-30 is able to increase the size exclusion limit from 700 Daltons to 9400 Daltons thereby aiding its movement through a plant. Also, increasing calcium concentrations in the cytoplasm, either by injection or by cold-induction, has been shown to constrict the opening of surrounding plasmodesmata and limit transport.

Several models for possible active transport through plasmodesmata exist. It has been suggested that such transport is mediated by interactions with proteins localized on the desmotubule, and/or by chaperones partially unfolding proteins, allowing them to fit through the narrow passage. A similar mechanism may be involved in transporting viral nucleic acids through the plasmodesmata.

References

- Brás, Natércia (2008). "Carbohydrate Binding Modules from family 11: Understanding the binding mode of polysaccharides". International Journal of Quantum Chemistry. 108 (11): 2030–2040. doi:10.1002/qua.21755
- Hobgood Ray, Kathryn (August 25, 2011). "Cars Could Run on Recycled Newspaper, Tulane Scientists Say". Tulane University news webpage. Tulane University. Retrieved March 14, 2012
- Buchanan; Gruissem, Jones (2000). Biochemistry & molecular biology of plants (1st ed.). American society of plant physiology. ISBN 0-943088-39-9
- Sendbusch, Peter V. (2003-07-31). "Cell Walls of Algae Archived November 28, 2005, at the Wayback Machine.". Botany Online. Retrieved on 2007-10-29

- Yablokov, Alexey V. Chernobyl Consequences of the Catastrophe for People and the Environment. John Wiley & Sons, 2010, pp. 304–309 ISBN 1573317578
- Richmond, Todd A; Somerville, Chris R (October 2000). "The Cellulose Synthase Superfamily". Plant Physiology. 124 (2): 495–498. PMC 1539280. PMID 11027699. doi:10.1104/pp.124.2.495. Retrieved 14 December 2014
- W. Boerjan; J. Ralph; M. Baucher (June 2003). "Lignin biosynthesis". Annu. Rev. Plant Biol. 54 (1): 519–549. PMID 14503002. doi:10.1146/annurev.arplant.54.031902.134938
- Sriamornsak, Pornsak (2003). "Chemistry of Pectin and its Pharmaceutical Uses: A Review". Silpakorn University International Journal. 3 (1–2): 206
- Lodish, Berk, Zipursky, Matsudaira, Baltimore, Darnell (2000). "22". Molecular Cell Biology (4 ed.). p. 998. ISBN 0-7167-3706-X. OCLC 41266312
- Stenius, Per (2000). "Ch. 1". Forest Products Chemistry. Papermaking Science and Technology. Vol. 3. Finland: Fapet OY. p. 35. ISBN 952-5216-03-9
- "Lignin and its Properties: Glossary of Lignin Nomenclature". Dialogue/Newsletters Volume 9, Number 1. Lignin Institute. July 2001. Retrieved 2007-10-14
- "Endogenous production of methanol after the consumption of fruit.". Alcohol Clin Exp Res. 21: 939–43. Aug 1997. PMID 9267548. doi:10.1111/j.1530-0277.1997.tb03862.x
- Gibson LJ (2013). "The hierarchical structure and mechanics of plant materials". Journal of the Royal Society Interface. 9 (76): 2749–2766. PMC 3479918. PMID 22874093. doi:10.1098/rsif.2012.0341

Chapter 5

Plastids and its Types

Plastids are the membrane bound organelles which are mainly responsible for the manufacturing and storage of food in plants. They also contain pigments used in photosynthesis. The types of pigments of the plastid are responsible for determining the cell color. All the diverse principles of plastids have been carefully analyzed in this chapter.

Plastid

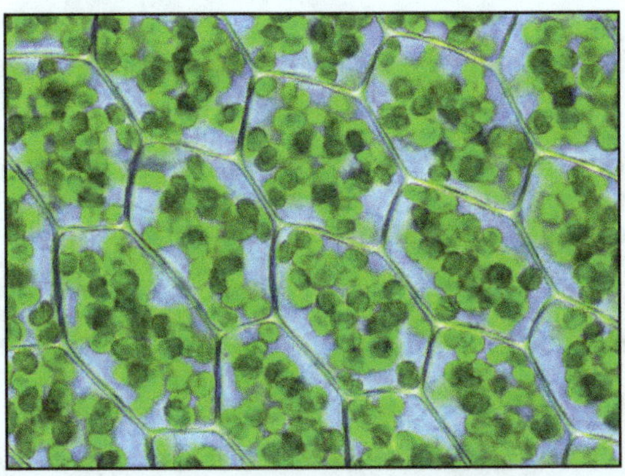

Plant cells with visible chloroplasts

The plastid is a major double-membrane organelle found, among others, in the cells of plants and algae. Plastids are the site of manufacture and storage of important chemical compounds used by the cell. They often contain pigments used in photosynthesis, and the types of pigments present can change or determine the cell's color. They have a common evolutionary origin and possess a double-stranded DNA molecule that is circular, like that of prokaryotic cells.

Plastids in Plants

Those plastids that contain chlorophyll can carry out photosynthesis. Plastids can also store products like starch and can synthesize fatty acids and terpenes, which can be used for producing energy and as raw material for the synthesis of other molecules. For example, the components of the plant cuticle and its epicuticular wax are synthesized by the epidermal cells from palmitic acid, which is synthesized in the chloroplasts of the mesophyll tissue. All plastids are derived from proplastids, which are present in the meristematic regions of the plant. Proplastids and young chloroplasts commonly divide by binary fission, but more mature chloroplasts also have this capacity.

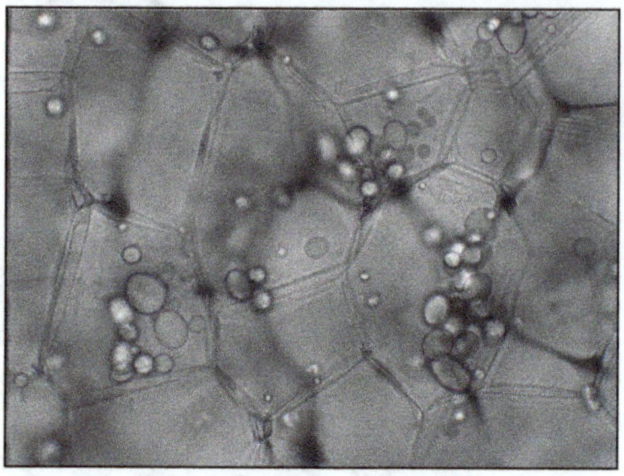

Leucoplasts in plant cells.

In plants, plastids may differentiate into several forms, depending upon which function they play in the cell. Undifferentiated plastids (*proplastids*) may develop into any of the following variants:

- Chloroplasts green plastids: for photosynthesis; *the predecessors of chloroplasts.*
- Chromoplasts coloured plastids: for pigment synthesis and storage.
- Gerontoplasts: control the dismantling of the photosynthetic apparatus during plant senescence.
- Leucoplasts colourless plastids: for monoterpene synthesis; *leucoplasts sometimes differentiate into more specialized plastids:*
    - Amyloplasts: for starch storage and detecting gravity.
    - Elaioplasts: for storing fat.
    - Proteinoplasts: for storing and modifying protein.
    - Tannosomes: for synthesizing and producing tannins and polyphenols.

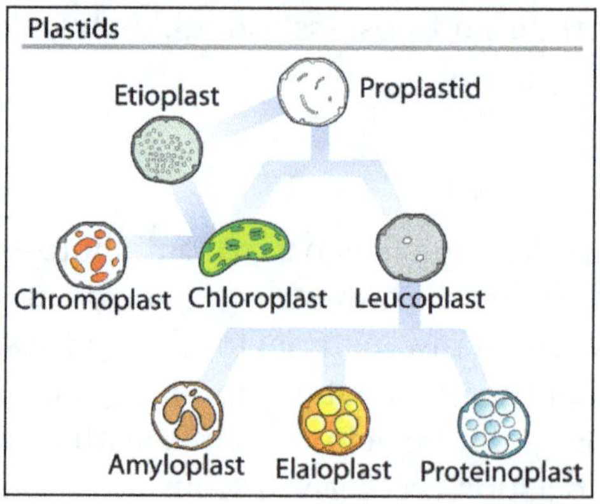

Depending on their morphology and function, plastids have the ability to differentiate, or redifferentiate, between these and other forms.

Each plastid creates multiple copies of a circular 75–250 kilobaseplastome. The number of genome copies per plastid is variable, ranging from more than 1000 in rapidly dividing cells, which, in general, contain few plastids, to 100 or fewer in mature cells, where plastid divisions have given rise to a large number of plastids. The plastome contains about 100 genes encoding ribosomal and transfer ribonucleic acids (rRNAs and tRNAs) as well as proteins involved in photosynthesis and plastid gene transcription and translation. However, these proteins only represent a small fraction of the total protein set-up necessary to build and maintain the structure and function of a particular type of plastid. Plant nuclear genes encode the vast majority of plastid proteins, and the expression of plastid genes and nuclear genes is tightly co-regulated to coordinate proper development of plastids in relation to cell differentiation.

Plastid DNA exists as large protein-DNA complexes associated with the inner envelope membrane and called 'plastid nucleoids'. Each nucleoid particle may contain more than 10 copies of the plastid DNA. The proplastid contains a single nucleoid located in the centre of the plastid. The developing plastid has many nucleoids, localized at the periphery of the plastid, bound to the inner envelope membrane. During the development of proplastids to chloroplasts, and when plastids convert from one type to another, nucleoids change in morphology, size and location within the organelle. The remodelling of nucleoids is believed to occur by modifications to the composition and abundance of nucleoid proteins.

Many plastids, particularly those responsible for photosynthesis, possess numerous internal membrane layers.

In plant cells, long thin protuberances called stromules sometimes form and extend from the main plastid body into the cytosol and interconnect several plastids. Proteins, and presumably smaller molecules, can move within stromules. Most cultured cells that are relatively large compared to other plant cells have very long and abundant stromules that extend to the cell periphery.

In 2014, evidence of possible plastid genome loss was found in *Rafflesia lagascae,* a non-photosynthetic parasitic flowering plant, and in *Polytomella,* a genus of non-photosynthetic green algae. Extensive searches for plastid genes in both *Rafflesia* and *Polytomella* yielded no results, however the conclusion that their plastomes are entirely missing is still controversial. Some scientists argue that plastid genome loss is unlikely since even non-photosynthetic plastids contain genes necessary to complete various biosynthetic pathways, such as heme biosynthesis.

Plastids in Algae

In algae, the term leucoplast is used for all unpigmented plastids and their function differs from the leucoplasts of plants. Etioplasts, amyloplasts and chromoplasts are plant-specific and do not occur in algae. Plastids in algae and hornworts may also differ from plant plastids in that they contain pyrenoids.

Glaucocystophytic algae contain muroplasts, which are similar to chloroplasts except that they have a peptidoglycan cell wall that is similar to that of prokaryotes. Rhodophytic algae contain rhodoplasts, which are red chloroplasts that allow the algae to photosynthesise to a depth of up to 268 m. The chloroplasts of plants differ from the rhodoplasts of red algae in their ability to synthesize starch, which is stored in the form of granules within the plastids. In red algae, floridean starch is synthesized and stored outside the plastids in the cytosol.

Inheritance of Plastids

Most plants inherit the plastids from only one parent. In general, angiosperms inherit plastids from the female gamete, whereas many gymnosperms inherit plastids from the male pollen. Algae also inherit plastids from only one parent. The plastid DNA of the other parent is, thus, completely lost.

In normal intraspecific crossings (resulting in normal hybrids of one species), the inheritance of plastid DNA appears to be quite strictly 100% uniparental. In interspecific hybridisations, however, the inheritance of plastids appears to be more erratic. Although plastids inherit mainly maternally in interspecific hybridisations, there are many reports of hybrids of flowering plants that contain plastids of the father. Approximately 20% of angiosperms, including alfalfa (*Medicago sativa*), normally show biparental inheritance of plastids.

Origin of Plastids

Plastids are thought to have originated from endosymbioticcyanobacteria. This symbiosis evolved around 1.5 billion years ago and enabled eukaryotes to carry out oxygenic photosynthesis. Three evolutionary lineages have since emerged in which the plastids are named differently: chloroplasts in green algae and plants, rhodoplasts in red algae and muroplasts in the glaucophytes. The plastids differ both in their pigmentation and in their ultrastructure. For example, chloroplasts have lost all phycobilisomes, the light harvesting complexes found in cyanobacteria, red algae and glaucophytes, but instead contain stroma and grana thylakoids, structures found only in plants and closely related green algae. The glaucocystophycean plastid — in contrast to chloroplasts and rhodoplasts — is still surrounded by the remains of the cyanobacterial cell wall. All these primary plastids are surrounded by two membranes.

Complex plastids start by secondary endosymbiosis (where a eukaryotic organism engulfs another eukaryotic organism that contains a primary plastid resulting in its endosymbiotic fixation), when a eukaryote engulfs a red or green alga and retains the algal plastid, which is typically surrounded by more than two membranes. In some cases these plastids may be reduced in their metabolic and/or photosynthetic capacity. Algae with complex plastids derived by secondary endosymbiosis of a red alga include the heterokonts, haptophytes, cryptomonads, and most dinoflagellates (= rhodoplasts). Those that endosymbiosed a green alga include the euglenids and chlorarachniophytes (= chloroplasts). The Apicomplexa, a phylum of obligate parasitic protozoa including the causative agents of malaria (*Plasmodium* spp.), toxoplasmosis (*Toxoplasma gondii*), and many other human or animal diseases also harbor a complex plastid (although this organelle has been lost in some apicomplexans, such as *Cryptosporidium parvum*, which causes cryptosporidiosis). The 'apicoplast' is no longer capable of photosynthesis, but is an essential organelle, and a promising target for antiparasitic drug development.

Some dinoflagellates and sea slugs, in particular of the genus *Elysia*, take up algae as food and keep the plastid of the digested alga to profit from the photosynthesis; after a while, the plastids are also digested. This process is known as kleptoplasty, from the Greek, *kleptes*, thief.

Chloroplast

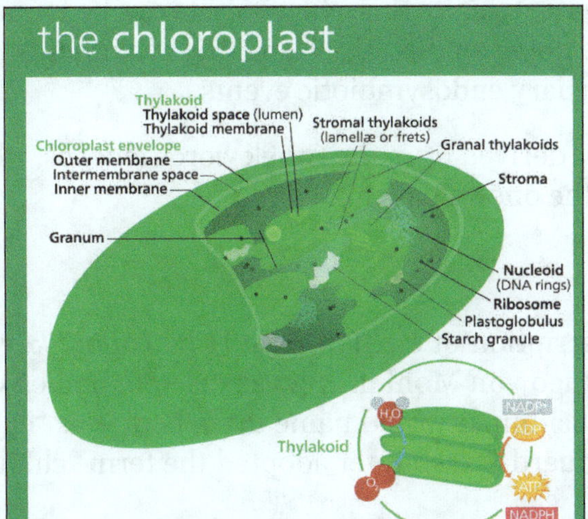

Structure of a typical higher-plant chloroplast

Chloroplasts are organelles, specialized subunits, in plant and algal cells. Their discovery inside plant cells is usually credited to Julius von Sachs (1832–1897), an influential botanist and author of standard botanical textbooks – sometimes called "The Father of Plant Physiology".

The main role of chloroplasts is to conduct photosynthesis, where the photosynthetic pigment chlorophyll captures the energy from sunlight and converts it and stores it in the energy-storage molecules ATP and NADPH while freeing oxygen from water. They then use the ATP and NADPH to make organic molecules from carbon dioxide in a process known as the Calvin cycle. Chloroplasts carry out a number of other functions, including fatty acid synthesis, much amino acid synthesis, and the immune response in plants. The number of chloroplasts per cell varies from one, in unicellular algae, up to 100 in plants like *Arabidopsis* and wheat.

A chloroplast is a type of organelle known as a plastid, characterized by its high concentration of chlorophyll. Other plastid types, such as the leucoplast and the chromoplast, contain little chlorophyll and do not carry out photosynthesis.

Chloroplasts are highly dynamic—they circulate and are moved around within plant cells, and occasionally pinch in two to reproduce. Their behavior is strongly influenced by environmental factors like light color and intensity. Chloroplasts, like mitochondria, contain their own DNA, which is thought to be inherited from their ancestor—a photosynthetic cyanobacterium that was engulfed by an early eukaryotic cell. Chloroplasts cannot be made by the plant cell and must be inherited by each daughter cell during cell division.

With one exception (the amoeboid*Paulinella chromatophora*), all chloroplasts can probably be traced back to a single endosymbiotic event, when a cyanobacterium was engulfed by the eukaryote. Despite this, chloroplasts can be found in an extremely wide set of organisms, some not even directly related to each other—a consequence of many secondary and even tertiary endosymbiotic events.

The word *chloroplast* is derived from the Greek words *chloros*, which means green, and *plastes*, which means the one who forms.

Discovery

The first definitive description of a chloroplast (*Chlorophyllkörnen*, "grain of chlorophyll") was given by Hugo von Mohl in 1837 as discrete bodies within the green plant cell. In 1883, A. F. W. Schimper would name these bodies as "chloroplastids" (*Chloroplastiden*). In 1884, Eduard Strasburger adopted the term "chloroplasts" (*Chloroplasten*).

Chloroplast Lineages and Evolution

Chloroplasts are one of many types of organelles in the plant cell. They are considered to have originated from cyanobacteria through endosymbiosis—when a eukaryotic cell engulfed a photosynthesizing cyanobacterium that became a permanent resident in the cell. Mitochondria are thought to have come from a similar event, where an aerobicprokaryote was engulfed. This origin of chloroplasts was first suggested by the Russian biologist Konstantin Mereschkowski in 1905 after Andreas Schimper observed in 1883 that chloroplasts closely resemble cyanobacteria. Chloroplasts are only found in plants, algae, and the amoeboid*Paulinella chromatophora*.

Cyanobacterial Ancestor

Cyanobacteria are considered the ancestors of chloroplasts. They are sometimes called blue-green algae even though they are prokaryotes. They are a diverse phylum of bacteria capable of carrying out photosynthesis, and are gram-negative, meaning that they have two cell membranes. Cyanobacteria also contain a peptidoglycan cell wall, which is thicker than in other gram-negative bacteria, and which is located between their two cell membranes. Like chloroplasts, they have thylakoids within. On the thylakoid membranes are photosynthetic pigments, including chlorophyll *a*.Phycobilins are also common cyanobacterial pigments, usually organized into hemispherical phycobilisomes attached to the outside of the thylakoid membranes (phycobilins are not shared with all chloroplasts though).

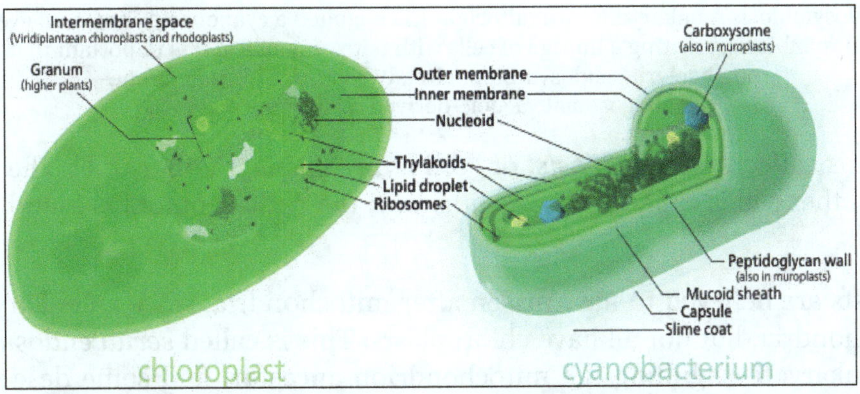

Both chloroplasts and cyanobacteria have a double membrane, DNA, ribosomes, and thylakoids. Both the chloroplast and cyanobacterium depicted are idealized versions (the chloroplast is that of a higher plant)—a lot of diversity exists among chloroplasts and cyanobacteria.

Primary Endosymbiosis

Somewhere around a billion years ago, a free-living cyanobacterium entered an early eukaryotic cell, either as food or as an internal parasite, but managed to escape the phagocytic vacuole it was contained in. The two innermost lipid-bilayer membranes that surround all chloroplasts correspond to the outer and inner membranes of the ancestral cyanobacterium's gram negative cell wall, and not the phagosomal membrane from the host, which was probably lost. The new cellular resident quickly became an advantage, providing food for the eukaryotic host, which allowed it to live within it. Over time, the cyanobacterium was assimilated, and many of its genes were lost or transferred to the nucleus of the host. From genomes that probably originally contained over 3000 genes only about 130 genes remain in the chloroplasts of contemporary plants. Some of its proteins were then synthesized in the cytoplasm of the host cell, and imported back into the chloroplast (formerly the cyanobacterium). Separately, somewhere around 100 million years ago, it happened again and led to the amoeboid-*Paulinella chromatophora*.

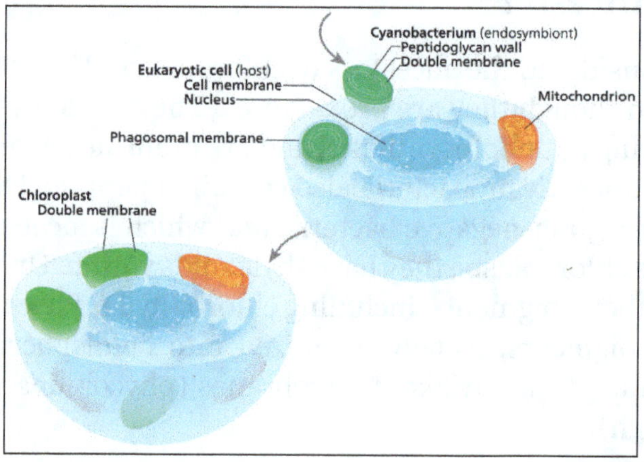

Primary endosymbiosis A eukaryote with mitochondria engulfed a cyanobacterium in an event of serial primary endosymbiosis, creating a lineage of cells with both organelles. It is important to note that the cyanobacterial endosymbiont already had a double membrane— the phagosomal vacuole-derived membrane was lost.

This event is called *endosymbiosis*, or "cell living inside another cell". The cell living inside the other cell is called the *endosymbiont*; the endosymbiont is found inside the *host cell.*

Chloroplasts are believed to have arisen after mitochondria, since all eukaryotes contain mitochondria, but not all have chloroplasts. This is called *serial endosymbiosis*— an early eukaryote engulfing the mitochondrion ancestor, and some descendants of it then engulfing the chloroplast ancestor, creating a cell with both chloroplasts and mitochondria.

Whether or not chloroplasts came from a single endosymbiotic event, or many independent engulfments across various eukaryotic lineages, has been long debated. It is now generally held that most organisms with chloroplasts either share a single ancestor or obtained their chloroplast from organisms that share a common ancestor that took in a cyanobacterium 600–1600 million years ago. The exception is the amoeboid*Paulinella chromatophora*, which descends from an ancestor that took in a cyanobacterium 90–140 million years ago.

These chloroplasts, which can be traced back directly to a cyanobacterial ancestor, are known as *primary plastids* (*"plastid"* in this context means almost the same thing as chloroplast). All primary chloroplasts belong to one of four chloroplast lineages—the glaucophyte chloroplast lineage, the amoeboid*Paulinella chromatophora* lineage, the rhodophyte (red algal) chloroplast lineage, or the chloroplastidan (green) chloroplast lineage. The rhodophyte and chloroplastidan lineages are the largest, with chloroplastidan (green) being the one that contains the land plants.

Primary endosymbiosis	Secondary endosymbiosis	Tertiary endosymbiosis
Glaucophyta		**Chloroplast lineages** A primary endosymbiosis event gave rise to four main lineages of chloroplasts in the glaucophytes, Paulinella, chlorophyta, and rhodophyta. Some of these algae were subsequently engulfed by other algae, becoming secondary (or tertiary) endosymbionts. [a] The apicomplexans (malaria parasites), contain a red algal endosymbiont with a non-photosynthetic chloroplast. [b] 2–3 chloroplast membranes [a c] 2–4 chloroplast membranes
Paulinella		
Chloroplastida Land plants Green algae	Euglenophyta Chlorarachniophyta Green algal dinophytes	
Rhodophyceae (Red algae)	Apicomplexa[a] Peridinin-typedinophytes [b]	
	Cryptophyta	
	Haptophyta	Haptophyte dinophytes[c]
	Heterokontophyta	Diatom dinophytes

Glaucophyta

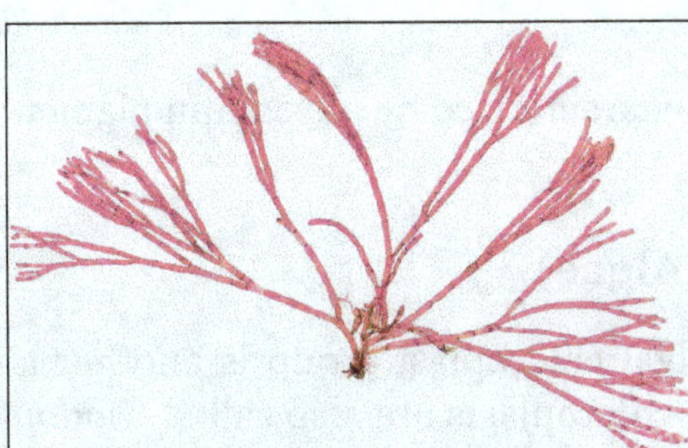

Bornetia secundiflora

The alga *Cyanophora*, a glaucophyte, is thought to be one of the first organisms to contain a chloroplast. The glaucophyte chloroplast group is the smallest of the three primary chloroplast lineages, being found in only 13 species, and is thought to be the one that branched off the earliest. Glaucophytes have chloroplasts that retain a peptidoglycan wall between their double membranes, like their cyanobacterial parent. For this reason, glaucophyte chloroplasts are also known as *muroplasts*. Glaucophyte chloroplasts also contain concentric unstacked thylakoids, which surround a carboxysome – an icosahedral structure that glaucophyte chloroplasts and cyanobacteria keep their carbon fixationenzymerubisco in. The starch that they synthesize collects outside the chloroplast. Like cyanobacteria, glaucophyte chloroplast thylakoids are studded with light collecting structures called phycobilisomes. For these reasons, glaucophyte chloroplasts are considered a primitive intermediate between cyanobacteria and the more evolved chloroplasts in red algae and plants.

Peyssonnelia squamaria.

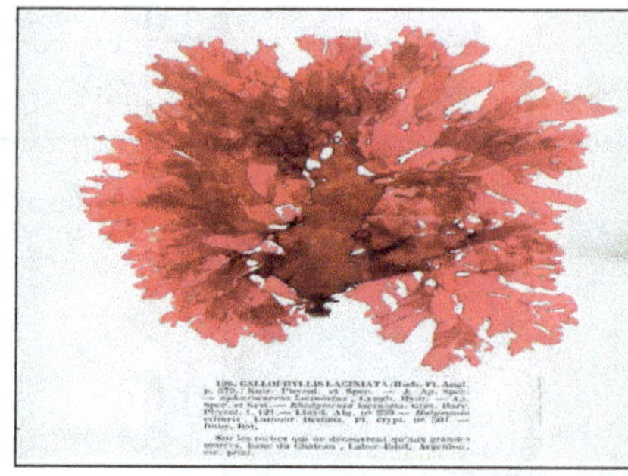

Cyanidium.

Laurencia.

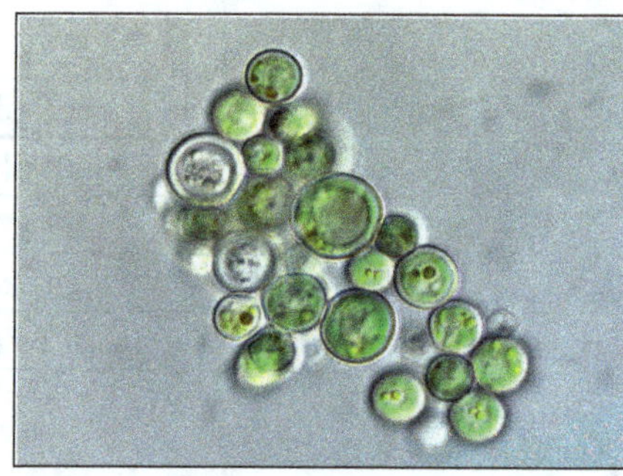

Callophyllis laciniata.

Red algal chloroplasts are characterized by phycobilin pigments which often give them their reddish color.

Rhodophyceae (Red Algae)

The rhodophyte, or red algal chloroplast group is another large and diverse chloroplast lineage. Rhodophyte chloroplasts are also called *rhodoplasts*, literally "red chloroplasts".

Rhodoplasts have a double membrane with an intermembrane space and phycobilin pigments organized into phycobilisomes on the thylakoid membranes, preventing their thylakoids from stacking. Some contain pyrenoids. Rhodoplasts have chlorophyll *a* and phycobilins for photosynthetic pigments; the phycobilin phycoerytherin is responsible for giving many red algae their distinctive red color. However, since they also contain the blue-green chlorophyll *a* and other pigments, many are reddish to purple from the combination. The red phycoerytherin pigment is an adaptation to help red algae catch more sunlight in deep water—as such, some red algae that live in shallow water have less phycoerytherin in their rhodoplasts, and can appear more greenish. Rhodoplasts synthesize a form of starch called floridean starch, which collects into granules outside the rhodoplast, in the cytoplasm of the red alga.

Chloroplastida (Green Algae and Plants)

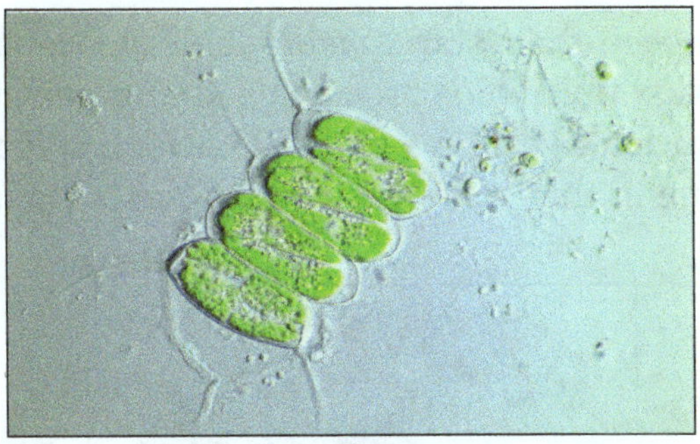

Scenedesmus

The chloroplastidan chloroplasts, or green chloroplasts, are another large, highly diverse primary chloroplast lineage. Their host organisms are commonly known as the green algae and land plants. They differ from glaucophyte and red algal chloroplasts in that they have lost their phycobilisomes, and contain chlorophyll *b* instead. Most green chloroplasts are (obviously) green, though some aren't, like some forms of *Hæmatococcus pluvialis*, due to accessory pigments that override the chlorophylls' green colors. Chloroplastidan chloroplasts have lost the peptidoglycan wall between their double membrane, leaving an intermembrane space. Some plants seem to have kept the genes for the synthesis of the peptidoglycan layer, though they've been repurposed for use in chloroplast division instead.

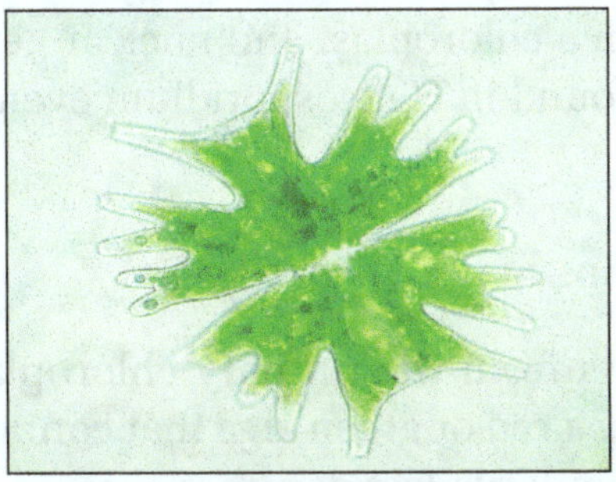

Micrasterias

Hydrodictyon

Volvox

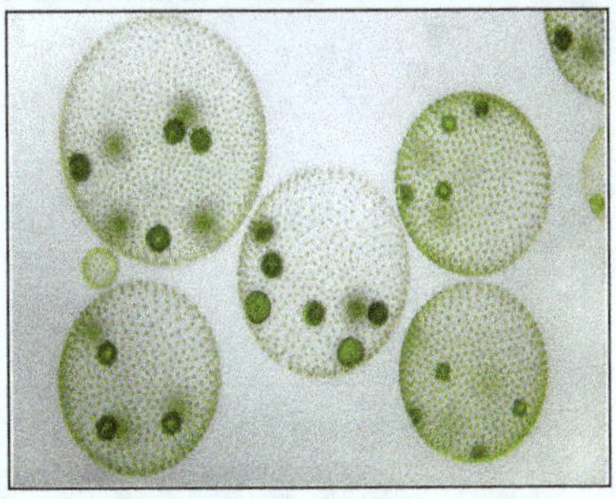

Stigeoclonium

Green algal chloroplasts are characterized by their pigments chlorophyll *a* and chlorophyll *b* which give them their green color.

Most of the chloroplasts depicted are green chloroplasts.

Green algae and plants keep their starch*inside* their chloroplasts, and in plants and some algae, the chloroplast thylakoids are arranged in grana stacks. Some green algal chloroplasts contain a structure called a pyrenoid, which is functionally similar to the glaucophyte carboxysome in that it is where rubisco and CO_2 are concentrated in the chloroplast.

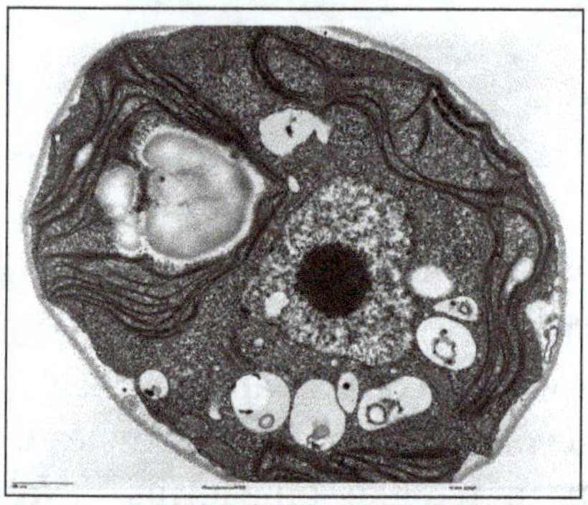

Transmission electron micrograph of *Chlamydomonas reinhardtii*, a green alga that contains a pyrenoid surrounded by starch.

Helicosporidium

Helicosporidium is a genus of nonphotosynthetic parasitic green algae that is thought to contain a vestigial chloroplast. Genes from a chloroplast and nuclear genes indicating the presence of a chloroplast have been found in Helicosporidium even if nobody's seen the chloroplast itself.

Secondary and Tertiary Endosymbiosis

Many other organisms obtained chloroplasts from the primary chloroplast lineages through secondary endosymbiosis—engulfing a red or green alga that contained a chloroplast. These chloroplasts are known as secondary plastids.

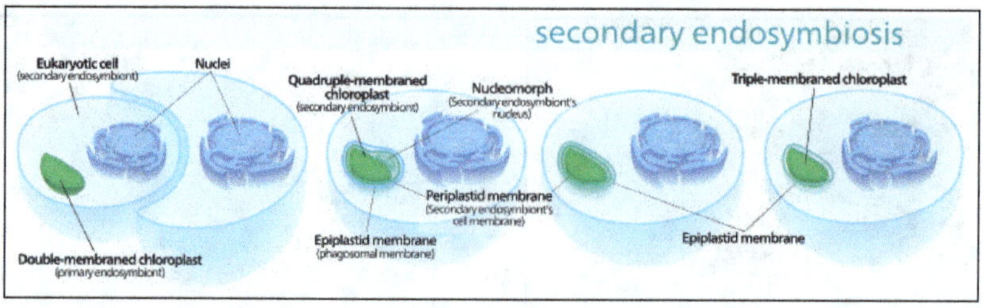

Secondary endosymbiosis consisted of a eukaryoticalga being engulfed by another eukaryote, forming a chloroplast with three or four membranes.

While primary chloroplasts have a double membrane from their cyanobacterial ancestor, secondary chloroplasts have additional membranes outside of the original two, as a result of the secondary endosymbiotic event, when a nonphotosynthetic eukaryote engulfed a chloroplast-containing alga but failed to digest it—much like the cyanobacterium at the beginning of this story. The engulfed alga was broken down, leaving only its chloroplast, and sometimes its cell membrane and nucleus, forming a chloroplast with three or four membranes—the two cyanobacterial membranes, sometimes the eaten alga's cell membrane, and the phagosomal vacuole from the host's cell membrane.

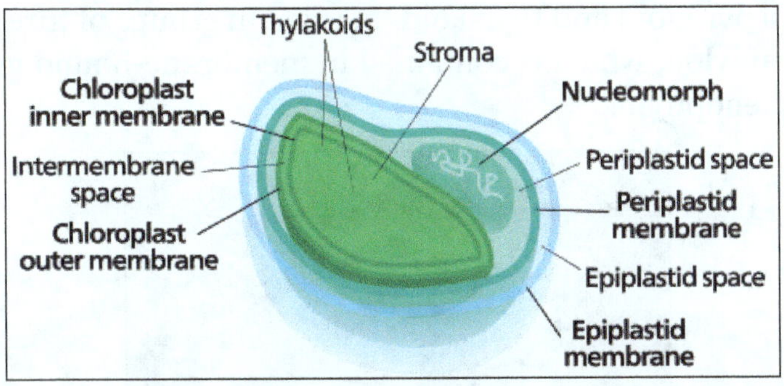

Diagram of a four membraned chloroplast containing a nucleomorph

The genes in the phagocytosed eukaryote's nucleus are often transferred to the secondary host's nucleus.Cryptomonads and chlorarachniophytes retain the phagocytosed eukaryote's nucleus, an object called a nucleomorph, located between the second and third membranes of the chloroplast.

All secondary chloroplasts come from green and red algae—no secondary chloroplasts from glaucophytes have been observed, probably because glaucophytes are relatively rare in nature, making them less likely to have been taken up by another eukaryote.

Green Algal Derived Chloroplasts

Green algae have been taken up by the euglenids, chlorarachniophytes, a lineage of dinoflagellates, and possibly the ancestor of the chromalveolates in three or four separate engulfments. Many green algal derived chloroplasts contain pyrenoids, but unlike chloroplasts in their green algal ancestors, starch collects in granules outside the chloroplast.

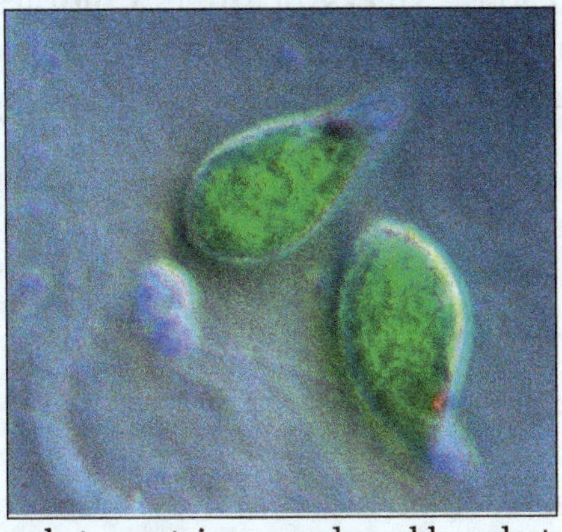

Euglena, a euglenophyte, contains secondary chloroplasts from green algae

Euglenophytes

Euglenophytes are a group of common flagellatedprotists that contain chloroplasts derived from a green alga.Euglenophyte chloroplasts have three membranes—it is thought that the membrane of the primary endosymbiont was lost, leaving the cyanobacterial membranes, and the secondary host's phagosomal membrane. Euglenophyte chloroplasts have a pyrenoid and thylakoids stacked in groups of three. Starch is stored in the form of paramylon, which is contained in membrane-bound granules in the cytoplasm of the euglenophyte.

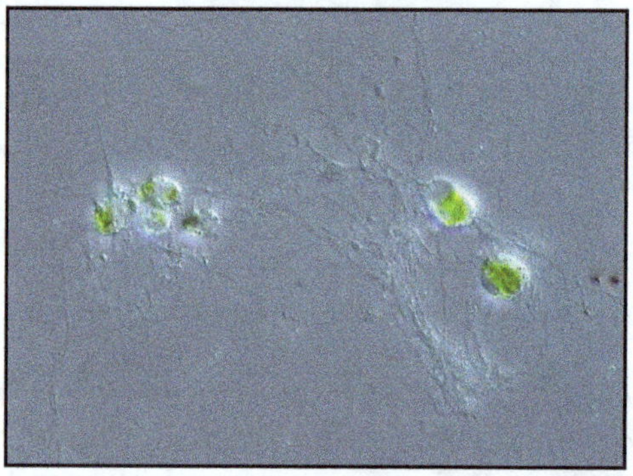

Chlorarachnion reptans is a chlorarachniophyte. Chlorarachniophytes replaced their original red algal endosymbiont with a green alga.

Chlorarachniophytes

Chlorarachniophytes are a rare group of organisms that also contain chloroplasts derived from green algae, though their story is more complicated than that of the euglenophytes. The ancestor of chlorarachniophytes is thought to have been a chromalveolate, a eukaryote with a *red* algal derived chloroplast. It is then thought to have lost its first red algal chloroplast, and later engulfed a green alga, giving it its second, green algal derived chloroplast.

Chlorarachniophyte chloroplasts are bounded by four membranes, except near the cell membrane, where the chloroplast membranes fuse into a double membrane. Their thylakoids are arranged in loose stacks of three. Chlorarachniophytes have a form of starch called chrysolaminarin, which they store in the cytoplasm, often collected around the chloroplast pyrenoid, which bulges into the cytoplasm.

Chlorarachniophyte chloroplasts are notable because the green alga they are derived from has not been completely broken down—its nucleus still persists as a nucleomorph found between the second and third chloroplast membranes—the periplastid space, which corresponds to the green alga's cytoplasm.

Early Chromalveolates

Recent research has suggested that the ancestor of the chromalveolates acquired a green algal prasinophyte endosymbiont. The green algal derived chloroplast was lost and replaced with a red algal derived chloroplast, but not before contributing some of its genes to the early chromalveolate's nucleus. The presence of both green algal and red algal genes in chromalveolates probably helps them thrive under fluctuating light conditions.

Red Algal Derived Chloroplasts (Chromalveolate Chloroplasts)

Like green algae, red algae have also been taken up in secondary endosymbiosis, though it is thought that all red algal derived chloroplasts are descended from a single red alga that was engulfed by an early chromalveolate, giving rise to the chromalveolates, some of which, like the ciliates, subsequently lost the chloroplast. This is still debated though.

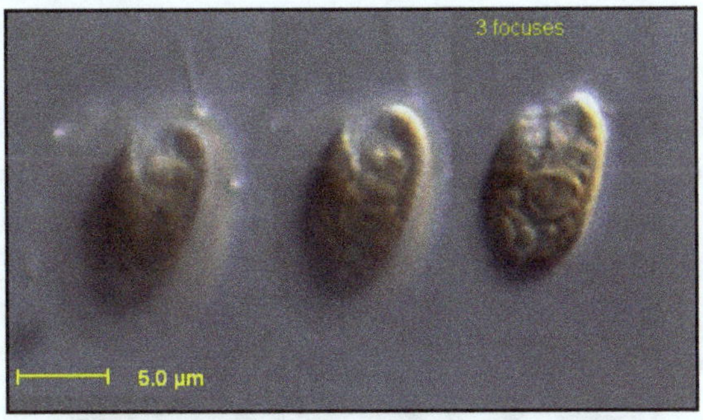

Rhodomonas salina is a cryptophyte

Pyrenoids and stacked thylakoids are common in chromalveolate chloroplasts, and the outermost membrane of many are continuous with the rough endoplasmic reticulum and studded with ribosomes. They have lost their phycobilisomes and exchanged them for chlorophyll *c*, which isn't found in primary red algal chloroplasts themselves.

Cryptophytes

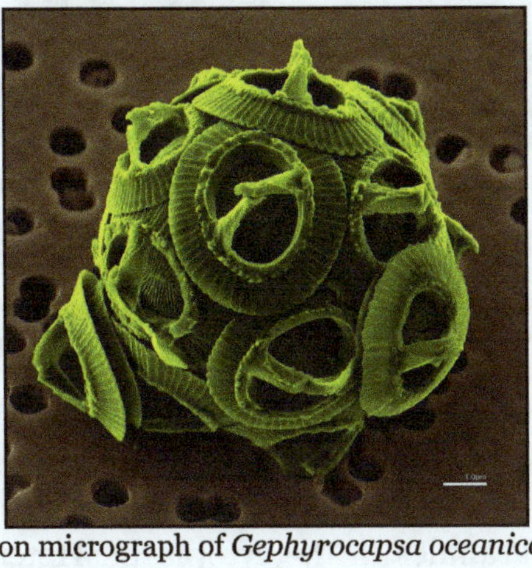

Scanning electron micrograph of *Gephyrocapsa oceanica*, a haptophyte.

Cryptophytes, or cryptomonads are a group of algae that contain a red-algal derived chloroplast. Cryptophyte chloroplasts contain a nucleomorph that superficially resembles that of the chlorarachniophytes. Cryptophyte chloroplasts have four membranes, the outermost of which is continuous with the rough endoplasmic reticulum. They synthesize ordinary starch, which is stored in granules found in the periplastid space—outside the original double membrane, in the place that corresponds to the red alga's cytoplasm. Inside cryptophyte chloroplasts is a pyrenoid and thylakoids in stacks of two.

Their chloroplasts do not have phycobilisomes, but they do have phycobilin pigments which they keep in their thylakoid space, rather than anchored on the outside of their thylakoid membranes.

Haptophytes

Haptophytes are similar and closely related to cryptophytes, and are thought to be the first chromalveolates to branch off. Their chloroplasts lack a nucleomorph, their thylakoids are in stacks of three, and they synthesize chrysolaminarin sugar, which they store completely outside of the chloroplast, in the cytoplasm of the haptophyte.

Heterokontophytes (Stramenopiles)

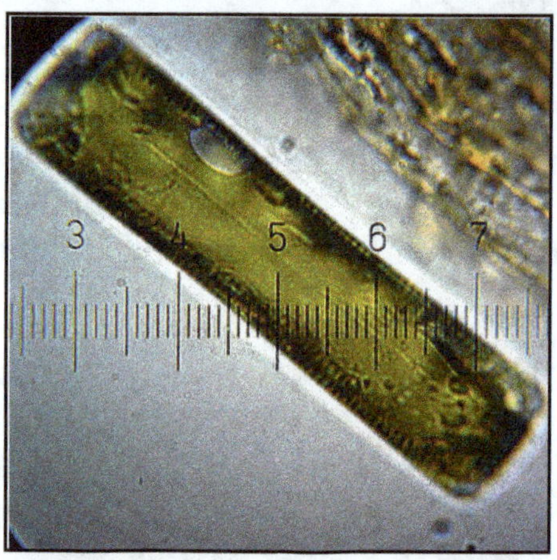

The photosynthetic pigments present in their chloroplasts give diatoms a greenish-brown color

The heterokontophytes, also known as the stramenopiles, are a very large and diverse group of algae that also contain red algal derived chloroplasts. Heterokonts include the diatoms and the brown algae, golden algae, and yellow-green algae.

Heterokont chloroplasts are very similar to haptophyte chloroplasts, containing a pyrenoid, triplet thylakoids, and with some exceptions, having an epiplastid membrane connected to the endoplasmic reticulum. Like haptophytes, heterokontophytes store sugar in chrysolaminarin granules in the cytoplasm. Heterokontophyte chloroplasts contain chlorophyll *a* and with a few exceptionschlorophyll *c*, but also have carotenoids which give them their many colors.

Apicomplexans

Apicomplexans are another group of chromalveolates. Like the helicosproidia, they're parasitic, and have a nonphotosynthetic chloroplast. They were once thought to be related to the helicosproidia, but it is now known that the helicosproida are green algae rather than chromalveolates. The apicomplexans include *Plasmodium*, the malaria parasite. Many apicomplexans keep a vestigial red algal derived chloroplast called an apicoplast, which they inherited from their ancestors. Other apicomplexans like *Cryptosporidium* have lost the chloroplast completely. Apicomplexans store their energy in amylopectin starch granules that are located in their cytoplasm, even though they are nonphotosynthetic.

Apicoplasts have lost all photosynthetic function, and contain no photosynthetic pigments or true thylakoids. They are bounded by four membranes, but the membranes are not connected to the endoplasmic reticulum. The fact that apicomplexans still keep their nonphotosynthetic chloroplast around demonstrates how the chloroplast carries out important functions other than photosynthesis. Plant chloroplasts provide plant cells with many important things besides sugar, and apicoplasts are no different—they synthesize fatty acids, isopentenyl pyrophosphate, iron-sulfur clusters, and carry out part of the heme pathway. This makes the apicoplast an attractive target for drugs to cure apicomplexan-related diseases.The most important apicoplast function is isopentenyl pyrophosphate synthesis—in fact, apicomplexans die when something interferes with this apicoplast function, and when apicomplexans are grown in an isopentenyl pyrophosphate-rich medium, they dump the organelle.

Dinophytes

The dinoflagellates are yet another very large and diverse group of protists, around half of which are (at least partially) photosynthetic.

Most dinophyte chloroplasts are secondary red algal derived chloroplasts, like other chromalveolate chloroplasts. Many other dinophytes have lost the chloroplast (becoming the nonphotosynthetic kind of dinoflagellate), or replaced it though *tertiary*endosymbiosis—the engulfment of another chromalveolate containing a red algal derived chloroplast. Others replaced their original chloroplast with a green algal derived one.

Most dinophyte chloroplasts contain at least the photosynthetic pigmentschlorophyll *a*, chlorophyll c_2, *beta*-carotene, and at least one dinophyte-unique xanthophyll (peridinin, dinoxanthin, or diadinoxanthin), giving many a golden-brown color. All dinophytes store starch in their cytoplasm, and most have chloroplasts with thylakoids arranged in stacks of three.

Peridinin-containing Dinophyte Chloroplast

The most common dinophyte chloroplast is the peridinin-type chloroplast, characterized by the carotenoid pigment peridinin in their chloroplasts, along with chlorophyll *a* and chlorophyll c_2. Peridinin is not found in any other group of chloroplasts. The peridinin chloroplast is bounded by three membranes (occasionally two), having lost the red algal endosymbiont's original cell membrane. The outermost membrane is not connected to the endoplasmic reticulum. They contain a pyrenoid, and have triplet-stacked thylakoids. Starch is found outside the chloroplast. An important feature of these chloroplasts is that their chloroplast DNA is highly reduced and fragmented into many small circles. Most of the genome has migrated to the nucleus, and only critical photosynthesis-related genes remain in the chloroplast.

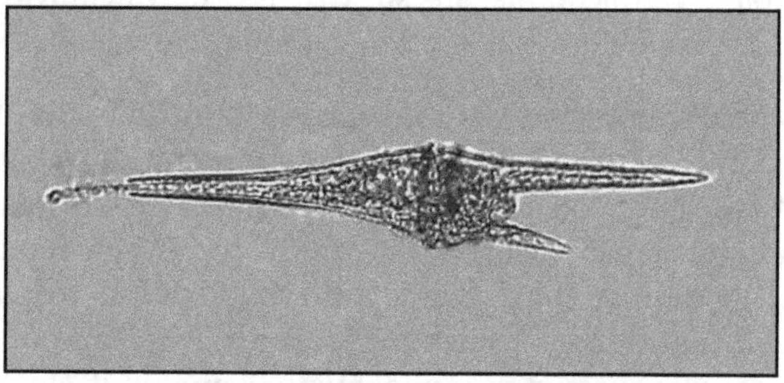

Ceratium furca, a peridinin-containing dinophyte

The peridinin chloroplast is thought to be the dinophytes' "original" chloroplast, which has been lost, reduced, replaced, or has company in several other dinophyte lineages.

Fucoxanthin-containing Dinophyte Chloroplasts (Haptophyte Endosymbionts)

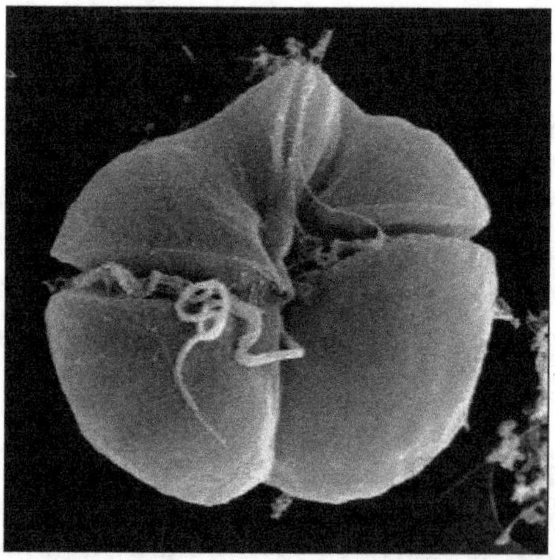

Karenia brevis is a fucoxanthin-containing dynophyte responsible for algal blooms called "red tides"

The fucoxanthin dinophyte lineages (including *Karlodinium* and *Karenia*) lost their original red algal derived chloroplast, and replaced it with a new chloroplast derived from a haptophyte endosymbiont. *Karlodinium* and *Karenia* probably took up different heterokontophytes. Because the haptophyte chloroplast has four membranes, tertiary endosymbiosis would be expected to create a six membraned chloroplast, adding the haptophyte's cell membrane and the dinophyte's phagosomal vacuole. However, the haptophyte was heavily reduced, stripped of a few membranes and its nucleus, leaving only its chloroplast (with its original double membrane), and possibly one or two additional membranes around it.

Fucoxanthin-containing chloroplasts are characterized by having the pigment fucoxanthin (actually 19′-hexanoyloxy-fucoxanthin and/or 19′-butanoyloxy-fucoxanthin) and no peridinin. Fucoxanthin is also found in haptophyte chloroplasts, providing evidence of ancestry.

Dinophysis acuminata has chloroplasts taken from a cryptophyte

Cryptophyte Derived Dinophyte Chloroplast

Members of the genus *Dinophysis* have a phycobilin-containing chloroplast taken from a cryptophyte. However, the cryptophyte is not an endosymbiont—only the chloroplast seems to have been taken, and the chloroplast has been stripped of its nucleomorph and outermost two membranes, leaving just a two-membraned chloroplast. Cryptophyte chloroplasts require their nucleomorph to maintain themselves, and *Dinophysis* species grown in cell culture alone cannot survive, so it is possible (but not confirmed) that the *Dinophysis* chloroplast is a kleptoplast—if so, *Dinophysis* chloroplasts wear out and *Dinophysis* species must continually engulf cryptophytes to obtain new chloroplasts to replace the old ones.

Diatom Derived Dinophyte Chloroplasts

Some dinophytes, like *Kryptoperidinium* and *Durinskia* have a diatom (heterokontophyte) derived chloroplast. These chloroplasts are bounded by up to *five* membranes, (depending on whether you count the entire diatom endosymbiont as the chloroplast, or just the red algal derived chloroplast inside it). The diatom endosymbiont has been reduced relatively little—it still retains its original mitochondria, and has endoplasmic reticulum, ribosomes, a nucleus, and of course, red algal derived chloroplasts—practically a complete cell, all inside the host's endoplasmic reticulum lumen. However the diatom endosymbiont can't store its own food—its starch is found in granules in the dinophyte host's cytoplasm instead. The diatom endosymbiont's nucleus is present, but it probably can't be called a nucleomorph because it shows no sign of genome reduction, and might have even been *expanded*. Diatoms have been engulfed by dinoflagellates at least three times.

The diatom endosymbiont is bounded by a single membrane, inside it are chloroplasts with four membranes. Like the diatom endosymbiont's diatom ancestor, the chloroplasts have triplet thylakoids and pyrenoids.

In some of these genera, the diatom endosymbiont's chloroplasts aren't the only chloroplasts in the dinophyte. The original three-membraned peridinin chloroplast is still around, converted to an eyespot.

Prasinophyte (Green Algal) Derived Dinophyte Chloroplast

Lepidodinium viride and its close relatives are dinophytes that lost their original peridinin chloroplast and replaced it with a green algal derived chloroplast (more specifically, a prasinophyte).*Lepidodinium* is the only dinophyte that has a chloroplast that's not from the rhodoplast lineage. The chloroplast is surrounded by two membranes and has no nucleomorph—all the nucleomorph genes have been transferred to the dinophyte nucleus. The endosymbiotic event that led to this chloroplast was serial secondary endosymbiosis rather than tertiary endosymbiosis—the endosymbiont was a green alga containing a primary chloroplast (making a secondary chloroplast).

Chromatophores

While most chloroplasts originate from that first set of endosymbiotic events, *Paulinella chromatophora* is an exception that acquired a photosynthetic cyanobacterial endosymbiont more recently. It is not clear whether that symbiont is closely related to the ancestral chloroplast of other eukaryotes. Being in the early stages of endosymbiosis, *Paulinella chromatophora* can offer some insights into how chloroplasts evolved.*Paulinella* cells contain one or two sausage shaped blue-green photosynthesizing structures called chromatophores, descended from the cyanobacterium *Synechococcus*. Chromatophores cannot survive outside their host. Chromatophore DNA is about a million base pairs long, containing around 850 protein encoding genes—far less than the three million base pair *Synechococcus* genome, but much larger than the approximately 150,000 base pair genome of the more assimilated chloroplast. Chromatophores have transferred much less of their DNA to the nucleus of their host. About 0.3–0.8% of the nuclear DNA in *Paulinella* is from the chromatophore, compared with 11–14% from the chloroplast in plants.

Kleptoplastidy

In some groups of mixotrophicprotists, like some dinoflagellates, chloroplasts are separated from a captured alga or diatom and used temporarily. These klepto chloroplasts may only have a lifetime of a few days and are then replaced.

Chloroplast DNA

Chloroplasts have their own DNA, often abbreviated as ctDNA, or cpDNA. It is also known as the plastome. Its existence was first proved in 1962, and first sequenced in 1986—when two Japanese research teams sequenced the chloroplast DNA of liverwort and tobacco. Since then, hundreds of chloroplast DNAs from various species have been sequenced, but they're mostly those of land plants and green algae—glaucophytes, red algae, and other algal groups are extremely underrepresented, potentially introducing some bias in views of "typical" chloroplast DNA structure and content.

Molecular Structure

With few exceptions, most chloroplasts have their entire chloroplast genome combined into a single large circular DNA molecule, typically 120,000–170,000 base pairs long. They can have a contour length of around 30–60 micrometers, and have a mass of about 80–130 million daltons.

While usually thought of as a circular molecule, there is some evidence that chloroplast DNA molecules more often take on a linear shape.

Inverted Repeats

Many chloroplast DNAs contain two *inverted repeats*, which separate a long single copy section (LSC) from a short single copy section (SSC). While a given pair of inverted repeats are rarely completely identical, they are always very similar to each other, apparently resulting from concerted evolution.

The inverted repeats vary wildly in length, ranging from 4,000 to 25,000 base pairs long each and containing as few as four or as many as over 150 genes. Inverted repeats in plants tend to be at the upper end of this range, each being 20,000–25,000 base pairs long.

The inverted repeat regions are highly conserved among land plants, and accumulate few mutations. Similar inverted repeats exist in the genomes of cyanobacteria and the other two chloroplast lineages (glaucophyta and rhodophyceae), suggesting that they predate the chloroplast, though some chloroplast DNAs have since lost or flipped the inverted repeats (making them direct repeats). It is possible that the inverted repeats help stabilize the rest of the chloroplast genome, as chloroplast DNAs which have lost some of the inverted repeat segments tend to get rearranged more.

Nucleoids

New chloroplasts may contain up to 100 copies of their DNA, though the number of chloroplast DNA copies decreases to about 15–20 as the chloroplasts age. They are usually packed into nucleoids, which can contain several identical chloroplast DNA rings. Many nucleoids can be found in each chloroplast. In primitive red algae, the chloroplast DNA nucleoids are clustered in the center of the chloroplast, while in green plants and green algae, the nucleoids are dispersed throughout the stroma.

Though chloroplast DNA is not associated with true histones, in red algae, similar proteins that tightly pack each chloroplast DNA ring into a nucleoid have been found.

DNA Replication

The Leading Model of cpDNA Replication

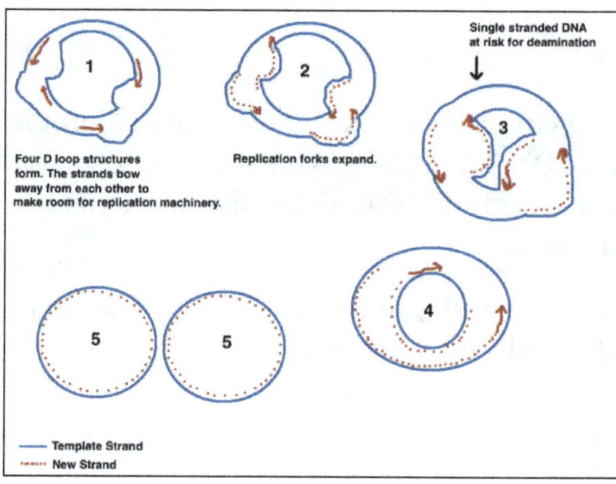

Chloroplast DNA replication via multiple D loop mechanisms.Adapted from Krishnan NM, Rao BJ's paper "A comparative approach to elucidate chloroplast genome replication."

The mechanism for chloroplast DNA (cpDNA) replication has not been conclusively determined, but two main models have been proposed. Scientists have attempted to observe chloroplast replication via electron microscopy since the 1970s. The results of the microscopy experiments led to the idea that chloroplast DNA replicates using a double displacement loop (D-loop). As the D-loop moves through the circular DNA, it adopts a theta intermediary form, also known as a Cairns replication intermediate, and completes replication with a rolling circle mechanism. Transcription starts at specific points of origin. Multiple replication forks open up, allowing replication machinery to transcribe the DNA. As replication continues, the forks grow and eventually converge. The new cpDNA structures separate, creating daughter cpDNA chromosomes.

In addition to the early microscopy experiments, this model is also supported by the amounts of deamination seen in cpDNA. Deamination occurs when an amino group is lost and is a mutation that often results in base changes. When adenine is deaminated, it becomes hypoxanthine. Hypoxanthine can bind to cytosine, and when the XC base pair is replicated, it becomes a GC (thus, an A → G base change).

Original DNA Strand

...CCATGCATGGATC...

Deamination of an Adenine

...CCATGCATGGATC...

↓

...CCHTGCATGGATC...

During Replication, H pairs with C

...CCHTGCATGGATC...
...GGCACGTACCTAG...

When Replicated Again, C pairs with G

...GGCACGTACCTAG...
...CCGTGCATGGATC...

Over time, base changes in the DNA sequence can arise from deamination mutations. When adenine is deaminated, it becomes hypoxanthine, which can pair with cytosine. During replication, the cytosine will pair with guanine, causing an A --> G base change.

Deamination

In cpDNA, there are several A → G deamination gradients. DNA becomes susceptible to deamination events when it is single stranded. When replication forks form, the strand not being copied is single stranded, and thus at risk for A → G deamination. Therefore, gradients in deamination indicate that replication forks were most likely present and the direction that they initially opened (the highest gradient is most likely nearest the start site because it was single stranded for the longest amount of time). This mechanism is still the leading theory today; however, a second theory suggests that most cpDNA is actually linear and replicates through homologous recombination. It further contends that only a minority of the genetic material is kept in circular chromosomes while the rest is in branched, linear, or other complex structures.

Alternative Model of Replication

One of competing model for cpDNA replication asserts that most cpDNA is linear and participates in homologous recombination and replication structures similar to bacteriophage T4. It has been established that some plants have linear cpDNA, such as maize, and that more species still contain complex structures that scientists do not yet understand. When the original experiments on cpDNA were performed, scientists did notice linear structures; however, they attributed these linear forms to broken circles. If the branched and complex structures seen in cpDNA experiments are real and not artifacts of concatenated circular DNA or broken circles, then a D-loop mechanism of replication is insufficient to explain how those structures would replicate. At the same time, homologous recombination does not expand the multiple A --> G gradients seen in plastomes. Because of the failure to explain the deamination gradient as well as the numerous plant species that have been shown to have circular cpDNA, the predominant theory continues to hold that most cpDNA is circular and most likely replicates via a D loop mechanism.

Gene Content and Protein Synthesis

The chloroplast genome most commonly includes around 100 genes that code for a variety of things, mostly to do with the protein pipeline and photosynthesis. As in prokaryotes, genes in chloroplast DNA are organized into operons. Interestingly though, unlike prokaryotic DNA molecules, chloroplast DNA molecules contain introns (plant mitochondrial DNAs do too, but not human mtDNAs).

Among land plants, the contents of the chloroplast genome are fairly similar.

Chloroplast Genome Reduction and Gene Transfer

Over time, many parts of the chloroplast genome were transferred to the nuclear genome of the host, a process called *endosymbiotic gene transfer*. As a result, the chloroplast genome is heavily reduced compared to that of free-living cyanobacteria. Chloroplasts may contain 60–100 genes whereas cyanobacteria often have more than 1500 genes in their genome.Recently, a plastid without a genome was found, demonstrating chloroplasts can lose their genome during endosymbiotic the gene transfer process.

Endosymbiotic gene transfer is how we know about the lost chloroplasts in many chromalveolate lineages. Even if a chloroplast is eventually lost, the genes it donated to the former host's nucleus persist, providing evidence for the lost chloroplast's existence. For example, while diatoms (a heterokontophyte) now have a red algal derived chloroplast, the presence of many green algal genes in the diatom nucleus provide evidence that the diatom ancestor (probably the ancestor of all chromalveolates too) had a green algal derived chloroplast at some point, which was subsequently replaced by the red chloroplast.

In land plants, some 11–14% of the DNA in their nuclei can be traced back to the chloroplast, up to 18% in *Arabidopsis*, corresponding to about 4,500 protein-coding genes. There have been a few recent transfers of genes from the chloroplast DNA to the nuclear genome in land plants.

Of the approximately 3000 proteins found in chloroplasts, some 95% of them are encoded by nuclear genes. Many of the chloroplast's protein complexes consist of subunits from both the chloroplast genome and the host's nuclear genome. As a result, protein synthesis must be coordinated between the chloroplast and the nucleus. The chloroplast is mostly under nuclear control, though chloroplasts can also give out signals regulating gene expression in the nucleus, called *retrograde signaling*.

Protein Synthesis

Protein synthesis within chloroplasts relies on two RNA polymerases. One is coded by the chloroplast DNA, the other is of nuclear origin. The two RNA polymerases may recognize and bind to different kinds of promoters within the chloroplast genome. The ribosomes in chloroplasts are similar to bacterial ribosomes.

Protein Targeting and Import

Because so many chloroplast genes have been moved to the nucleus, many proteins that would originally have been translated in the chloroplast are now synthesized in the cytoplasm of the plant cell. These proteins must be directed back to the chloroplast, and imported through at least two chloroplast membranes.

Curiously, around half of the protein products of transferred genes aren't even targeted back to the chloroplast. Many became exaptations, taking on new functions like participating in cell division, protein routing, and even disease resistance. A few chloroplast genes found new homes in the mitochondrial genome—most became nonfunctional pseudogenes, though a few tRNA genes still work in the mitochondrion. Some transferred chloroplast DNA protein products get directed to the secretory pathway though it should be noted that many secondary plastids are bounded by an outermost membrane derived from the host's cell membrane, and therefore topologically outside of the cell, because to reach the chloroplast from the cytosol, you have to cross the cell membrane, just like if you were headed for the extracellular space. In those cases, chloroplast-targeted proteins do initially travel along the secretory pathway.

Because the cell acquiring a chloroplast already had mitochondria (and peroxisomes, and a cell membrane for secretion), the new chloroplast host had to develop a unique protein targeting system to avoid having chloroplast proteins being sent to the wrong organelle.

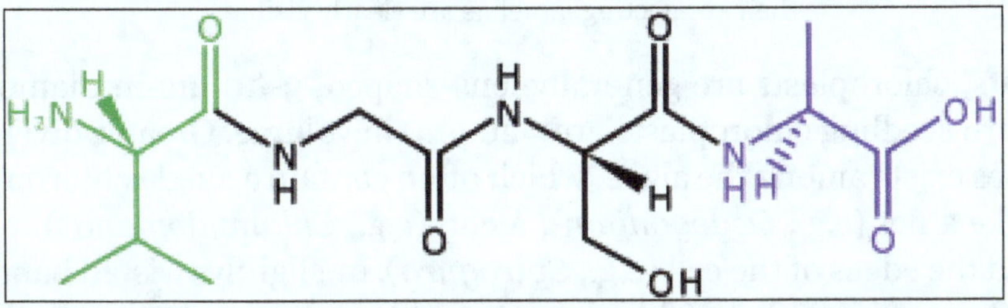

The two ends of a polypeptide are called the N-terminus, or *amino end*, and the C-terminus, or *carboxyl end*. This polypeptide has four amino acids linked together. At the left is the N-terminus, with its amino (H_2**N**) group in green. The blue C-terminus, with its carboxyl group (**CO**$_2$H) is at the right.

In most, but not all cases, nuclear-encoded chloroplast proteins are translated with a *cleavable transit peptide* that's added to the N-terminus of the protein precursor. Sometimes the transit sequence is found on the C-terminus of the protein, or within the functional part of the protein.

Transport Proteins and Membrane Translocons

After a chloroplast polypeptide is synthesized on a ribosome in the cytosol, an enzyme specific to chloroplast proteinsphosphorylates, or adds a phosphate group to many (but not all) of them in their transit sequences. Phosphorylation helps many proteins bind the polypeptide, keeping it from folding prematurely. This is important because it prevents chloroplast proteins from assuming their active form and carrying out their chloroplast functions in the wrong place—the cytosol. At the same time, they have to keep just enough shape so that they can be recognized by the chloroplast. These proteins also help the polypeptide get imported into the chloroplast.

From here, chloroplast proteins bound for the stroma must pass through two protein complexes—the TOC complex, or *translocon on the outer chloroplast membrane*, and the TIC translocon, or *translocon on the inner chloroplast membrane translocon*. Chloroplast polypeptide chains probably often travel through the two complexes at the same time, but the TIC complex can also retrieve preproteins lost in the intermembrane space.

Structure

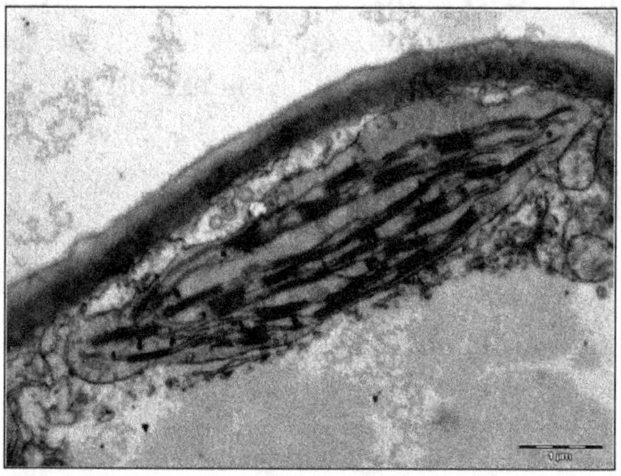

Transmission electron microscope image of a chloroplast. Grana of thylakoids and their connecting lamellae are clearly visible.

In land plants, chloroplasts are generally lens-shaped, 3–10 μm in diameter and 1–3 μm thick. Corn seedling chloroplasts are ≈20 μm^3 in volume. Greater diversity in chloroplast shapes exists among the algae, which often contain a single chloroplast that can be shaped like a net (e.g., *Oedogonium*), a cup (e.g., *Chlamydomonas*), a ribbon-like spiral around the edges of the cell (e.g., *Spirogyra*), or slightly twisted bands at the cell edges (e.g., *Sirogonium*). Some algae have two chloroplasts in each cell; they are star-shaped in *Zygnema*, or may follow the shape of half the cell in orderDesmidiales. In some algae, the chloroplast takes up most of the cell, with pockets for the nucleus and other organelles, for example, some species of *Chlorella* have a cup-shaped chloroplast that occupies much of the cell.

All chloroplasts have at least three membrane systems—the outer chloroplast membrane, the inner chloroplast membrane, and the thylakoid system. Chloroplasts that are the product of secondary endosymbiosis may have additional membranes surrounding these three. Inside the outer and inner chloroplast membranes is the chloroplast stroma, a semi-gel-like fluid that makes up much of a chloroplast's volume, and in which the thylakoid system floats.

There are some common misconceptions about the outer and inner chloroplast membranes. The fact that chloroplasts are surrounded by a double membrane is often cited as evidence that they are the descendants of endosymbiotic cyanobacteria. This is often interpreted as meaning the outer chloroplast membrane is the product of the host's cell membrane infolding to form a vesicle to surround the ancestral cyanobacterium—which is not true—both chloroplast membranes are homologous to the cyanobacterium's original double membranes.

The chloroplast double membrane is also often compared to the mitochondrial double membrane. This is not a valid comparison—the inner mitochondria membrane is used to run proton pumps and carry out oxidative phosphorylation across to generate ATP energy. The only chloroplast structure that can consideredanalogous to it is the internal thylakoid system. Even so, in terms of "in-out", the direction of chloroplast H^+ ion flow is in the opposite direction compared to oxidative phosphorylation in mitochondria. In addition, in terms of function, the inner chloroplast membrane, which regulates metabolite passage and synthesizes some materials, has no counterpart in the mitochondrion.

Outer Chloroplast Membrane

The outer chloroplast membrane is a semi-porous membrane that small molecules and ions can easily diffuse across. However, it is not permeable to larger proteins, so chloroplast polypeptides being synthesized in the cell cytoplasm must be transported across the outer chloroplast membrane by the TOC complex, or *translocon on the **o**uter **c**hloroplast* membrane.

The chloroplast membranes sometimes protrude out into the cytoplasm, forming a stromule, or stroma-containing tubule. Stromules are very rare in chloroplasts, and are much more common in other plastids like chromoplasts and amyloplasts in petals and roots, respectively. They may exist to increase the chloroplast's surface area for cross-membrane transport, because they are often branched and tangled with the endoplasmic reticulum. When they were first observed in 1962, some plant biologists dismissed the structures as artifactual, claiming that stromules were just oddly shaped chloroplasts with constricted regions or dividing chloroplasts. However, there is a growing body of evidence that stromules are functional, integral features of plant cell plastids, not merely artifacts.

Intermembrane Space and Peptidoglycan Wall

Usually, a thin intermembrane space about 10–20 nanometers thick exists between the outer and inner chloroplast membranes.

Instead of an intermembrane space, glaucophyte algae have a peptidoglycan wall between their inner and outer chloroplast membranes.

Glaucophyte algal chloroplasts have a peptidoglycan layer between the chloroplast membranes. It corresponds to the peptidoglycan cell wall of their cyanobacterial ancestors, which is located between their two cell membranes. These chloroplasts are called *muroplasts* (from Latin *"mura"*, meaning "wall"). Other chloroplasts have lost the cyanobacterial wall, leaving an intermembrane space between the two chloroplast envelope membranes.

Inner Chloroplast Membrane

The inner chloroplast membrane borders the stroma and regulates passage of materials in and out of the chloroplast. After passing through the TOC complex in the outer chloroplast membrane, polypeptides must pass through the TIC complex*(translocon on the inner chloroplast membrane)* which is located in the inner chloroplast membrane.

In addition to regulating the passage of materials, the inner chloroplast membrane is where fatty acids, lipids, and carotenoids are synthesized.

Peripheral Reticulum

Some chloroplasts contain a structure called the chloroplast peripheral reticulum. It is often found in the chloroplasts of C_4 plants, though it has also been found in some C_3 angiosperms, and even some gymnosperms. The chloroplast peripheral reticulum consists of a maze of membranous tubes and vesicles continuous with the inner chloroplast membrane that extends into the internal stromal fluid of the chloroplast. Its purpose is thought to be to increase the chloroplast's surface area for cross-membrane transport between its stroma and the cell cytoplasm. The small vesicles sometimes observed may serve as transport vesicles to shuttle stuff between the thylakoids and intermembrane space.

Stroma

The protein-rich,alkaline,aqueous fluid within the inner chloroplast membrane and outside of the thylakoid space is called the stroma, which corresponds to the cytosol of the original cyanobacterium. Nucleoids of chloroplast DNA, chloroplast ribosomes, the thylakoid system with plastoglobuli, starch granules, and many proteins can be found floating around in it. The Calvin cycle, which fixes CO_2 into sugar takes place in the stroma.

Chloroplast Ribosomes

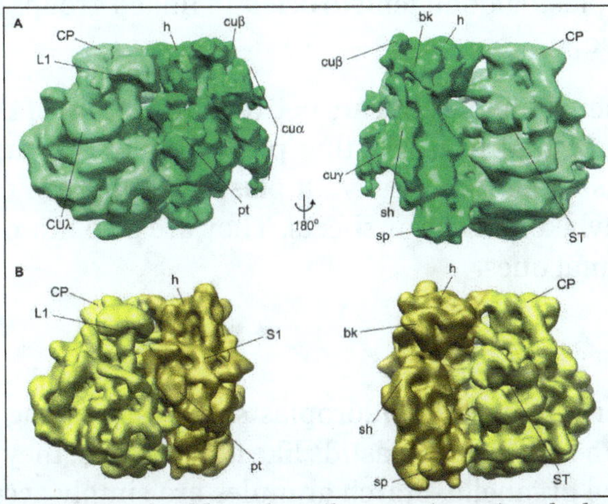

Chloroplast ribosomes Comparison of a chloroplast ribosome (green) and a bacterial ribosome (yellow). Important features common to both ribosomes and chloroplast-unique features are labeled.

Chloroplasts have their own ribosomes, which they use to synthesize a small fraction of their proteins. Chloroplast ribosomes are about two-thirds the size of cytoplasmic ribosomes (around 17 nm vs 25 nm). They take mRNAs transcribed from the chloroplast DNA and translate them into protein. While similar to bacterial ribosomes, chloroplast translation is more complex than in bacteria, so chloroplast ribosomes include some chloroplast-unique features. Small subunit ribosomal RNAs in several Chlorophyta and euglenid chloroplasts lack motifs for shine-dalgarno sequence recognition, which is considered essential for translation initiation in most chloroplasts and prokaryotes. Such loss is also rarely observed in other plastids and prokaryotes.

Plastoglobuli

Plastoglobuli (singular*plastoglobulus*, sometimes spelled *plastoglobule(s)*), are spherical bubbles of lipids and proteins about 45–60 nanometers across.They are surrounded by a lipid monolayer. Plastoglobuli are found in all chloroplasts, but become more common when the chloroplast is under oxidative stress, or when it ages and transitions into a gerontoplast. Plastoglobuli also exhibit a greater size variation under these conditions.They are also common in etioplasts, but decrease in number as the etioplasts mature into chloroplasts.

Plastoglubuli contain both structural proteins and enzymes involved in lipid synthesis and metabolism. They contain many types of lipids including plastoquinone, vitamin E, carotenoids and chlorophylls.

Plastoglobuli were once thought to be free-floating in the stroma, but it is now thought that they are permanently attached either to a thylakoid or to another plastoglobulus attached to a thylakoid, a configuration that allows a plastoglobulus to exchange its contents with the thylakoid network.In normal green chloroplasts, the vast majority of plastoglobuli occur singularly, attached directly to their parent thylakoid. In old or stressed chloroplasts, plastoglobuli tend to occur in linked groups or chains, still always anchored to a thylakoid.

Plastoglobuli form when a bubble appears between the layers of the lipid bilayer of the thylakoid membrane, or bud from existing plastoglubuli—though they never detach and float off into the stroma. Practically all plastoglobuli form on or near the highly curved edges of the thylakoid disks or sheets. They are also more common on stromal thylakoids than on granal ones.

Starch Granules

Starch granules are very common in chloroplasts, typically taking up 15% of the organelle's volume, though in some other plastids like amyloplasts, they can be big enough to distort the shape of the organelle. Starch granules are simply accumulations of starch in the stroma, and are not bounded by a membrane.

Starch granules appear and grow throughout the day, as the chloroplast synthesizes sugars, and are consumed at night to fuel respiration and continue sugar export into the phloem,though in mature chloroplasts, it is rare for a starch granule to be completely consumed or for a new granule to accumulate.

Starch granules vary in composition and location across different chloroplast lineages. In red algae, starch granules are found in the cytoplasm rather than in the chloroplast. In C_4 plants, mesophyll chloroplasts, which do not synthesize sugars, lack starch granules.

Rubisco

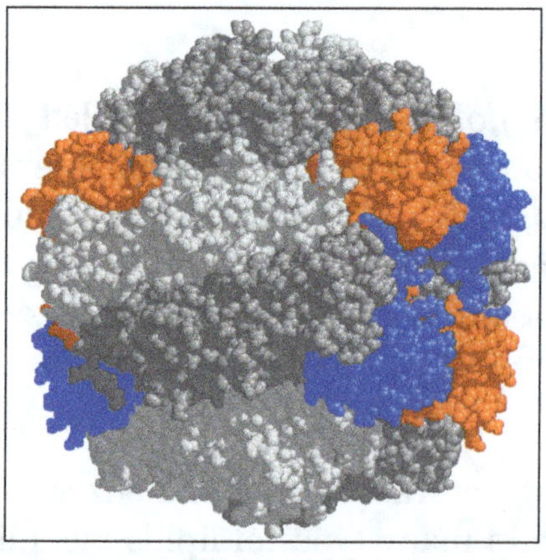

Rubisco, shown here in a space-filling model, is the main enzyme responsible for carbon fixation in chloroplasts.

The chloroplast stroma contains many proteins, though the most common and important is Rubisco, which is probably also the most abundant protein on the planet.Rubisco is the enzyme that fixes CO_2 into sugar molecules. In C_3 plants, rubisco is abundant in all chloroplasts, though in C_4 plants, it is confined to the bundle sheath chloroplasts, where the Calvin cycle is carried out in C_4 plants.

Pyrenoids

The chloroplasts of some hornworts and algae contain structures called pyrenoids. They are not found in higher plants. Pyrenoids are roughly spherical and highly refractive bodies which are a site of starch accumulation in plants that contain them. They consist of a matrix opaque to electrons, surrounded by two hemispherical starch plates. The starch is accumulated as the pyrenoids mature.In algae with carbon concentrating mechanisms, the enzyme rubisco is found in the pyrenoids. Starch can also accumulate around the pyrenoids when CO_2 is scarce. Pyrenoids can divide to form new pyrenoids, or be produced "de novo".

Thylakoid System

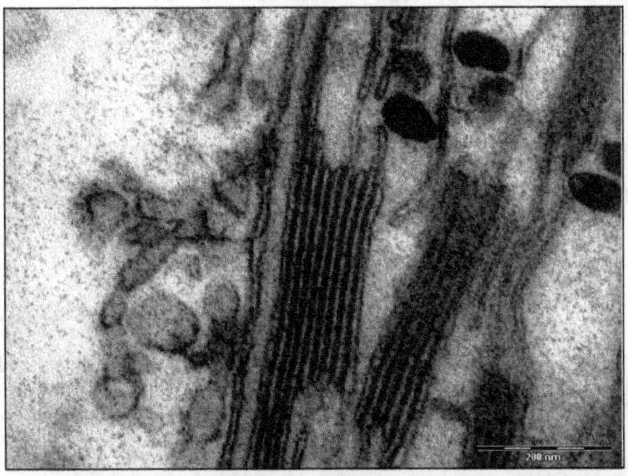

Transmission electron microscope image of some thylakoids arranged in grana stacks and lamellæ. Plastoglobuli (dark blobs) are also present.

Suspended within the chloroplast stroma is the thylakoid system, a highly dynamic collection of membranous sacks called thylakoids where chlorophyll is found and the light reactions of photosynthesis happen. In most vascular plant chloroplasts, the thylakoids are arranged in stacks called grana, though in certain C_4 plant chloroplasts and some algal chloroplasts, the thylakoids are free floating.

Granal Structure

Using a light microscope, it is just barely possible to see tiny green granules—which were named grana. With electron microscopy, it became possible to see the thylakoid system in more detail, revealing it to consist of stacks of flat thylakoids which made up the grana, and long interconnecting stromal thylakoids which linked different grana.

In the transmission electron microscope, thylakoid membranes appear as alternating light-and-dark bands, 8.5 nanometers thick.

For a long time, the three-dimensional structure of the thylakoid system has been unknown or disputed. One model has the granum as a stack of thylakoids linked by helical stromal thylakoids; the other has the granum as a single folded thylakoid connected in a "hub and spoke" way to other grana by stromal thylakoids. While the thylakoid system is still commonly depicted according to the folded thylakoid model, it was determined in 2011 that the stacked and helical thylakoids model is correct.

In the helical thylakoid model, grana consist of a stack of flattened circular granal thylakoids that resemble pancakes. Each granum can contain anywhere from two to a hundred thylakoids, though grana with 10–20 thylakoids are most common. Wrapped around the grana are helicoid stromal thylakoids, also known as frets or lamellar thylakoids. The helices ascend at an angle of 20–25°, connecting to each granal thylakoid at a bridge-like slit junction. The helicoids may extend as large sheets that link multiple grana, or narrow to tube-like bridges between grana.While different parts of the thylakoid system contain different membrane proteins, the thylakoid membranes are continuous and the thylakoid space they enclose form a single continuous labyrinth.

Thylakoids

Thylakoids (sometimes spelled *thylakoïds*), are small interconnected sacks which contain the membranes that the light reactions of photosynthesis take place on. The word *thylakoid* comes from the Greek word *thylakos* which means "sack".

Embedded in the thylakoid membranes are important protein complexes which carry out the light reactions of photosynthesis. Photosystem II and photosystem I contain light-harvesting complexes with chlorophyll and carotenoids that absorb light energy and use it to energize electrons. Molecules in the thylakoid membrane use the energized electrons to pump hydrogen ions into the thylakoid space, decreasing the pH and turning it acidic. ATP synthase is a large protein complex that harnesses the concentration gradient of the hydrogen ions in the thylakoid space to generate ATP energy as the hydrogen ions flow back out into the stroma—much like a dam turbine.

There are two types of thylakoids—granal thylakoids, which are arranged in grana, and stromal thylakoids, which are in contact with the stroma. Granal thylakoids are pancake-shaped circular disks about 300–600 nanometers in diameter. Stromal thylakoids are helicoid sheets that spiral around grana.The flat tops and bottoms of granal thylakoids contain only the relatively flat photosystem II protein complex. This allows them to stack tightly, forming grana with many layers of tightly appressed membrane, called granal membrane, increasing stability and surface area for light capture.

In contrast, photosystem I and ATP synthase are large protein complexes which jut out into the stroma. They can't fit in the appressed granal membranes, and so are found in the stromal thylakoid membrane—the edges of the granal thylakoid disks and the stromal thylakoids. These large protein complexes may act as spacers between the sheets of stromal thylakoids.

The number of thylakoids and the total thylakoid area of a chloroplast is influenced by light exposure. Shaded chloroplasts contain larger and more grana with more thylakoid membrane area than chloroplasts exposed to bright light, which have smaller and fewer grana and less thylakoid area. Thylakoid extent can change within minutes of light exposure or removal.

Pigments and Chloroplast Colors

Inside the photosystems embedded in chloroplast thylakoid membranes are various photosynthetic pigments, which absorb and transfer light energy. The types of pigments found are different in various groups of chloroplasts, and are responsible for a wide variety of chloroplast colorations.

Chlorophylls

Chlorophyll *a* is found in all chloroplasts, as well as their cyanobacterial ancestors. Chlorophyll *a* is a blue-greenpigment partially responsible for giving most cyanobacteria and chloroplasts their color. Other forms of chlorophyll exist, such as the accessory pigmentschlorophyll *b*, chlorophyll *c*, chlorophyll *d*, and chlorophyll *f*.

Chlorophyll *b* is an olive green pigment found only in the chloroplasts of plants, green algae, any secondary chloroplasts obtained through the secondary endosymbiosis of a green alga, and a few cyanobacteria. It is the chlorophylls *a* and *b* together that make most plant and green algal chloroplasts green.

Chlorophyll *c* is mainly found in secondary endosymbiotic chloroplasts that originated from a red alga, although it is not found in chloroplasts of red algae themselves. Chlorophyll *c* is also found in some green algae and cyanobacteria.

Chlorophylls *d* and *f* are pigments found only in some cyanobacteria.

Carotenoids

In addition to chlorophylls, another group of yellow–orange pigments called carotenoids are also found in the photosystems. There are about thirty photosynthetic carotenoids. They help transfer and dissipate excess energy, and their bright colors sometimes override the chlorophyll green, like during the fall, when the leaves of some land plants change color.β-carotene is a bright red-orange carotenoid found in nearly all chloroplasts, like chlorophyll *a*.Xanthophylls, especially the orange-red zeaxanthin, are also common.Many other forms of carotenoids exist that are only found in certain groups of chloroplasts.

Delesseria sanguinea, a red alga, has chloroplasts that contain red pigments like phycoerytherin that mask their blue-green chlorophyll *a*.

Phycobilins

Phycobilins are a third group of pigments found in cyanobacteria, and glaucophyte, red algal, and cryptophyte chloroplasts. Phycobilins come in all colors, though phycoerytherin is one of the pigments that makes many red algae red. Phycobilins often organize into relatively large protein complexes about 40 nanometers across called phycobilisomes. Like photosystem I and ATP synthase, phycobilisomes jut into the stroma, preventing thylakoid stacking in red algal chloroplasts.Cryptophyte chloroplasts and some cyanobacteria don't have their phycobilin pigments organized into phycobilisomes, and keep them in their thylakoid space instead.

Specialized Chloroplasts in C4 Plants

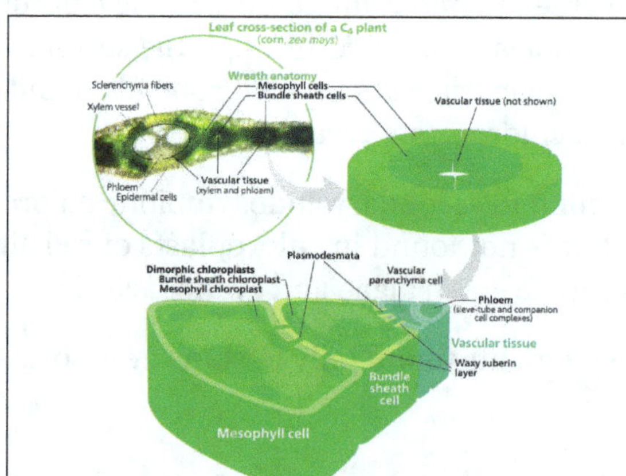

Many C_4 plants have their mesophyll cells and bundle sheath cells arranged radially around their leaf veins. The two types of cells contain different types of chloroplasts specialized for a particular part of photosynthesis.

To fix carbon dioxide into sugar molecules in the process of photosynthesis, chloroplasts use an enzyme called rubisco. Rubisco has a problem—it has trouble distinguishing between carbon dioxide and oxygen, so at high oxygen concentrations, rubisco starts accidentally adding oxygen to sugar precursors. This has the end result of ATP energy being wasted and CO_2 being released, all with no sugar being produced. This is a big problem, since O_2 is produced by the initial light reactions of photosynthesis, causing issues down the line in the Calvin cycle which uses rubisco.

C_4 plants evolved a way to solve this—by spatially separating the light reactions and the Calvin cycle. The light reactions, which store light energy in ATP and NADPH, are done in the mesophyll cells of a C_4 leaf. The Calvin cycle, which uses the stored energy to make sugar using rubisco, is done in the bundle sheath cells, a layer of cells surrounding a vein in a leaf.

As a result, chloroplasts in C_4 mesophyll cells and bundle sheath cells are specialized for each stage of photosynthesis. In mesophyll cells, chloroplasts are specialized for the light reactions, so they lack rubisco, and have normal grana and thylakoids, which they use to make ATP and NADPH, as well as oxygen. They store CO_2 in a four-carbon compound, which is why the process is called *C_4 photosynthesis*. The four-carbon compound is then transported to the bundle sheath chloroplasts, where it drops off CO_2 and returns to the mesophyll. Bundle sheath chloroplasts do not carry out the light reactions, preventing oxygen from building up in them and disrupting rubisco activity. Because of this, they lack thylakoids organized into grana stacks—though bundle sheath chloroplasts still have free-floating thylakoids in the stroma where they still carry out cyclic electron flow, a light-driven method of synthesizing ATP to power the Calvin cycle without generating oxygen. They lack photosystem II, and only have photosystem I—the only protein complex needed for cyclic electron flow. Because the job of bundle sheath chloroplasts is to carry out the Calvin cycle and make sugar, they often contain large starch grains.

Both types of chloroplast contain large amounts of chloroplast peripheral reticulum, which they use to get more surface area to transport stuff in and out of them. Mesophyll chloroplasts have a little more peripheral reticulum than bundle sheath chloroplasts.

Location

Distribution in a Plant

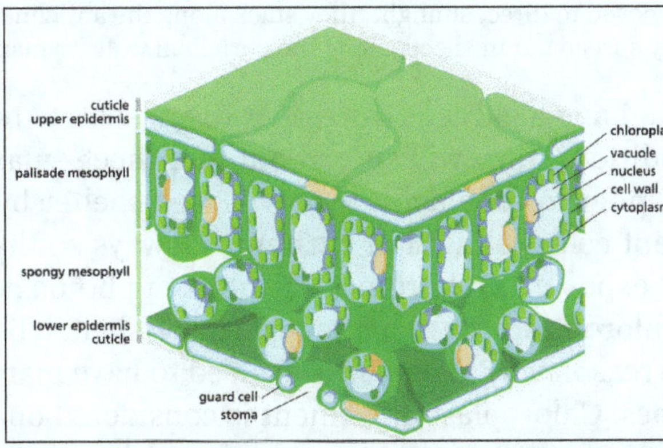

A cross section of a leaf, showing chloroplasts in its mesophyll cells. Stomal guard cells also have chloroplasts, though much fewer than mesophyll cells.

Not all cells in a multicellular plant contain chloroplasts. All green parts of a plant contain chloroplasts—the chloroplasts, or more specifically, the chlorophyll in them are what make the photosynthetic parts of a plant green. The plant cells which contain chloroplasts are usually parenchyma cells, though chloroplasts can also be found in collenchyma tissue.A plant cell which contains chloroplasts is known as a chlorenchyma cell. A typical chlorenchyma cell of a land plant contains about 10 to 100 chloroplasts.

In some plants such as cacti, chloroplasts are found in the stems, though in most plants, chloroplasts are concentrated in the leaves. One square millimeter of leaf tissue can contain half a million chloroplasts. Within a leaf, chloroplasts are mainly found in the mesophyll layers of a leaf, and the guard cells of stomata. Palisade mesophyll cells can contain 30–70 chloroplasts per cell, while stomatal guard cells contain only around 8–15 per cell, as well as much less chlorophyll. Chloroplasts can also be found in the bundle sheath cells of a leaf, especially in C_4 plants, which carry out the Calvin cycle in their bundle sheath cells. They are often absent from the epidermis of a leaf.

Cellular Location

Chloroplast Movement

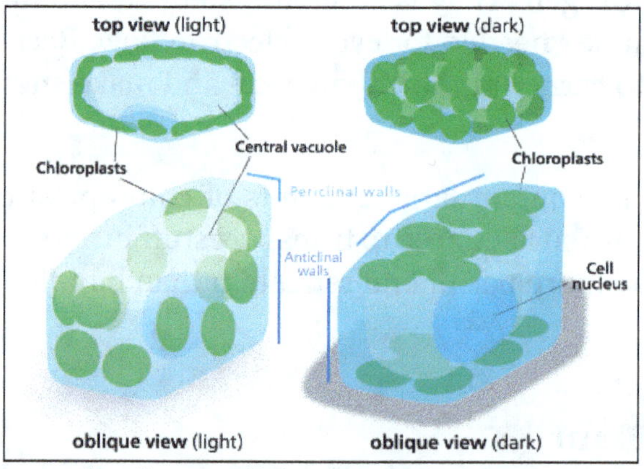

When chloroplasts are exposed to direct sunlight, they stack along the anticlinal cell walls to minimize exposure. In the dark they spread out in sheets along the periclinal walls to maximize light absorption.

The chloroplasts of plant and algal cells can orient themselves to best suit the available light. In low-light conditions, they will spread out in a sheet—maximizing the surface area to absorb light. Under intense light, they will seek shelter by aligning in vertical columns along the plant cell's cell wall or turning sideways so that light strikes them edge-on. This reduces exposure and protects them from photooxidative damage. This ability to distribute chloroplasts so that they can take shelter behind each other or spread out may be the reason why land plants evolved to have many small chloroplasts instead of a few big ones. Chloroplast movement is considered one of the most closely regulated stimulus-response systems that can be found in plants.Mitochondria have also been observed to follow chloroplasts as they move.

In higher plants, chloroplast movement is run by phototropins, blue light photoreceptors also responsible for plant phototropism. In some algae, mosses, ferns, and flowering plants, chloroplast movement is influenced by red light in addition to blue light, though very long red wavelengths inhibit movement rather than speeding it up. Blue light generally causes chloroplasts to seek shelter, while red light draws them out to maximize light absorption.

Studies of *Vallisneria gigantea*, an aquatic flowering plant, have shown that chloroplasts can get moving within five minutes of light exposure, though they don't initially show any net directionality. They may move along microfilament tracks, and the fact that the microfilament mesh changes shape to form a honeycomb structure surrounding the chloroplasts after they have moved suggests that microfilaments may help to anchor chloroplasts in place.

Function and Chemistry

Guard Cell Chloroplasts

Unlike most epidermal cells, the guard cells of plant stomata contain relatively well-developed chloroplasts. However, exactly what they do is controversial.

Plant Innate Immunity

Plants lack specialized immune cells—all plant cells participate in the plant immune response. Chloroplasts, along with the nucleus, cell membrane, and endoplasmic reticulum, are key players in pathogen defense. Due to its role in a plant cell's immune response, pathogens frequently target the chloroplast.

Plants have two main immune responses—the hypersensitive response, in which infected cells seal themselves off and undergo programmed cell death, and systemic acquired resistance, where infected cells release signals warning the rest of the plant of a pathogen's presence. Chloroplasts stimulate both responses by purposely damaging their photosynthetic system, producing reactive oxygen species. High levels of reactive oxygen species will cause the hypersensitive response. The reactive oxygen species also directly kill any pathogens within the cell. Lower levels of reactive oxygen species initiate systemic acquired resistance, triggering defense-molecule production in the rest of the plant.

In some plants, chloroplasts are known to move closer to the infection site and the nucleus during an infection.

Chloroplasts can serve as cellular sensors. After detecting stress in a cell, which might be due to a pathogen, chloroplasts begin producing molecules like salicylic acid, jasmonic acid, nitric oxide and reactive oxygen species which can serve as defense-signals. As cellular signals, reactive oxygen species are unstable molecules, so they probably don't leave the chloroplast, but instead pass on their signal to an unknown second messenger molecule. All these molecules initiate retrograde signaling—signals from the chloroplast that regulate gene expression in the nucleus.

In addition to defense signaling, chloroplasts, with the help of the peroxisomes, help synthesize an important defense molecule, jasmonate. Chloroplasts synthesize all the fatty acids in a plant cell—linoleic acid, a fatty acid, is a precursor to jasmonate.

Photosynthesis

One of the main functions of the chloroplast is its role in photosynthesis, the process by which light is transformed into chemical energy, to subsequently produce food in the form of sugars. Water (H_2O) and carbon dioxide (CO_2) are used in photosynthesis, and sugar and oxygen (O_2) is made, using light energy. Photosynthesis is divided into two stages—the light reactions, where water is split to produce oxygen, and the dark reactions, or Calvin cycle, which builds sugar molecules from carbon dioxide. The two phases are linked by the energy carriersadenosine triphosphate (ATP) and nicotinamide adenine dinucleotide phosphate ($NADP^+$).

Light Reactions

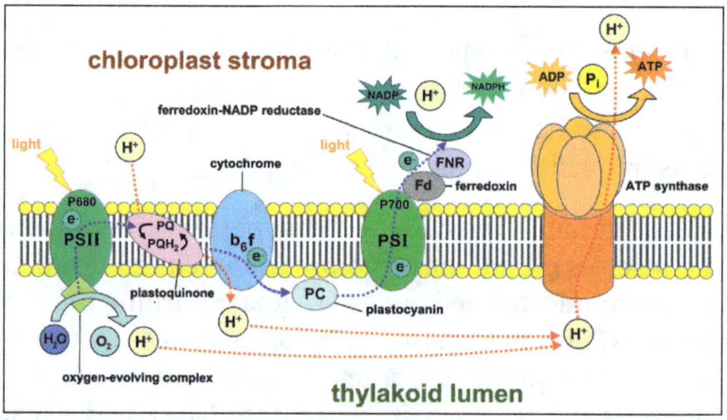

The light reactions of photosynthesis take place across the thylakoid membranes.

The light reactions take place on the thylakoid membranes. They take light energy and store it in NADPH, a form of $NADP^+$, and ATP to fuel the dark reactions.

Energy Carriers

ATP is the phosphorylated version of adenosine diphosphate (ADP), which stores energy in a cell and powers most cellular activities. ATP is the energized form, while ADP is the (partially) depleted form. $NADP^+$ is an electron carrier which ferries high energy electrons. In the light reactions, it gets reduced, meaning it picks up electrons, becoming NADPH.

Photophosphorylation

Like mitochondria, chloroplasts use the potential energy stored in an H^+, or hydrogen ion gradient to generate ATP energy. The two photosystems capture light energy to energize electrons taken from water, and release them down an electron transport chain. The molecules between the photosystems harness the electrons' energy to pump hydrogen ions into the thylakoid space, creating a concentration gradient, with more hydrogen ions (up to a thousand times as many) inside the thylakoid system than in the stroma. The hydrogen ions in the thylakoid space then diffuse back down their concentration gradient, flowing back out into the stroma through ATP synthase. ATP synthase uses the energy from the flowing hydrogen ions to phosphorylateadenosine diphosphate into adenosine triphosphate, or ATP. Because chloroplast ATP synthase projects out into the stroma, the ATP is synthesized there, in position to be used in the dark reactions.

$NADP^+$ Reduction

Electrons are often removed from the electron transport chains to charge $NADP^+$ with electrons, reducing it to NADPH. Like ATP synthase, ferredoxin-$NADP^+$ reductase, the enzyme that reduces $NADP^+$, releases the NADPH it makes into the stroma, right where it is needed for the dark reactions.

Because $NADP^+$ reduction removes electrons from the electron transport chains, they must be replaced—the job of photosystem II, which splits water molecules (H_2O) to obtain the electrons from its hydrogen atoms.

Cyclic Photophosphorylation

While photosystem IIphotolyzes water to obtain and energize new electrons, photosystem I simply reenergizes depleted electrons at the end of an electron transport chain. Normally, the reenergized electrons are taken by $NADP^+$, though sometimes they can flow back down more H^+-pumping electron transport chains to transport more hydrogen ions into the thylakoid space to generate more ATP. This is termed cyclic photophosphorylation because the electrons are recycled. Cyclic photophosphorylation is common in C_4 plants, which need more ATP than NADPH.

Dark Reactions

The Calvin cycle, also known as the dark reactions, is a series of biochemical reactions that fixes CO_2 into G3P sugar molecules and uses the energy and electrons from the ATP and NADPH made in the light reactions. The Calvin cycle takes place in the stroma of the chloroplast.

While named *"the dark reactions"*, in most plants, they take place in the light, since the dark reactions are dependent on the products of the light reactions.

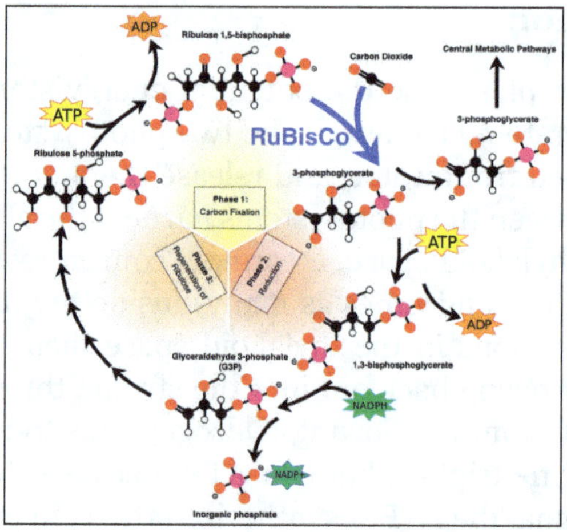

The Calvin cycle incorporates carbon dioxide into sugar molecules.

Carbon Fixation and G3P Synthesis

The Calvin cycle starts by using the enzyme Rubisco to fix CO_2 into five-carbon Ribulose bisphosphate (RuBP) molecules. The result is unstable six-carbon molecules that immediately break down into three-carbon molecules called 3-phosphoglyceric acid, or 3-PGA. The ATP and NADPH made in the light reactions is used to convert the 3-PGA into glyceraldehyde-3-phosphate, or G3P sugar molecules. Most of the G3P molecules are recycled back into RuBP using energy from more ATP, but one out of every six produced leaves the cycle—the end product of the dark reactions.

Sugars and Starches

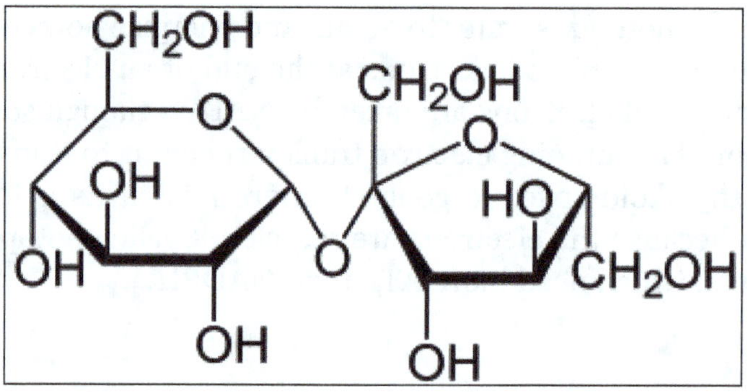

Sucrose is made up of a glucose monomer (left), and a fructose monomer (right)

Glyceraldehyde-3-phosphate can double up to form larger sugar molecules like glucose and fructose. These molecules are processed, and from them, the still larger sucrose, a disaccharide commonly known as table sugar, is made, though this process takes place outside of the chloroplast, in the cytoplasm.

Alternatively, glucose monomers in the chloroplast can be linked together to make starch, which accumulates into the starch grains found in the chloroplast.Under conditions such as high atmospheric CO_2 concentrations, these starch grains may grow very large, distorting the grana and thylakoids. The starch granules displace the thylakoids, but leave them intact. Waterlogged roots can also cause starch buildup in the chloroplasts, possibly due to less sucrose being exported out of the chloroplast (or more accurately, the plant cell). This depletes a plant's free phosphate supply, which indirectly stimulates chloroplast starch synthesis. While linked to low photosynthesis rates, the starch grains themselves may not necessarily interfere significantly with the efficiency of photosynthesis, and might simply be a side effect of another photosynthesis-depressing factor.

Photorespiration

Photorespiration can occur when the oxygen concentration is too high. Rubisco cannot distinguish between oxygen and carbon dioxide very well, so it can accidentally add O_2 instead of CO_2 to RuBP. This process reduces the efficiency of photosynthesis—it consumes ATP and oxygen, releases CO_2, and produces no sugar. It can waste up to half the carbon fixed by the Calvin cycle. Several mechanisms have evolved in different lineages that raise the carbon dioxide concentration relative to oxygen within the chloroplast, increasing the efficiency of photosynthesis. These mechanisms are called carbon dioxide concentrating mechanisms, or CCMs. These include Crassulacean acid metabolism, C_4 carbon fixation, and pyrenoids. Chloroplasts in C_4 plants are notable as they exhibit a distinct chloroplast dimorphism.

pH

Because of the H^+ gradient across the thylakoid membrane, the interior of the thylakoid is acidic, with a pH around 4, while the stroma is slightly basic, with a pH of around 8.The optimal stroma pH for the Calvin cycle is 8.1, with the reaction nearly stopping when the pH falls below 7.3.

CO_2 in water can form carbonic acid, which can disturb the pH of isolated chloroplasts, interfering with photosynthesis, even though CO_2 is used in photosynthesis. However, chloroplasts in living plant cells are not affected by this as much.

Chloroplasts can pump K^+ and H^+ ions in and out of themselves using a poorly understood light-driven transport system.

In the presence of light, the pH of the thylakoid lumen can drop up to 1.5 pH units, while the pH of the stroma can rise by nearly one pH unit.

Amino Acid Synthesis

Chloroplasts alone make almost all of a plant cell's amino acids in their stroma except the sulfur-containing ones like cysteine and methionine. Cysteine is made in the chloroplast (the proplastid too) but it is also synthesized in the cytosol and mitochondria, probably because it has trouble crossing membranes to get to where it is needed.The chloroplast is known to make the precursors to methionine but it is unclear whether the organelle carries out the last leg of the pathway or if it happens in the cytosol.

Other Nitrogen Compounds

Chloroplasts make all of a cell's purines and pyrimidines—the nitrogenous bases found in DNA and RNA. They also convert nitrite (NO_2^-) into ammonia (NH_3) which supplies the plant with nitrogen to make its amino acids and nucleotides.

Other Chemical Products

Chloroplasts are the site of complex lipid metabolism.

Differentiation, Replication, and Inheritance

Chloroplasts are a special type of a plant cell organelle called a plastid, though the two terms are sometimes used interchangeably. There are many other types of plastids, which carry out various functions. All chloroplasts in a plant are descended from undifferentiated proplastids found in the zygote, or fertilized egg. Proplastids are commonly found in an adult plant's apical meristems. Chloroplasts do not normally develop from proplastids in root tip meristems—instead, the formation of starch-storing amyloplasts is more common.

In shoots, proplastids from shoot apical meristems can gradually develop into chloroplasts in photosynthetic leaf tissues as the leaf matures, if exposed to the required light. This process involves invaginations of the inner plastid membrane, forming sheets of membrane that project into the internal stroma. These membrane sheets then fold to form thylakoids and grana.

If angiosperm shoots are not exposed to the required light for chloroplast formation, proplastids may develop into an etioplast stage before becoming chloroplasts. An etioplast is a plastid that lacks chlorophyll, and has inner membrane invaginations that form a lattice of tubes in their stroma, called a prolamellar body. While etioplasts lack chlorophyll, they have a yellow chlorophyll precursor stocked. Within a few minutes of light exposure, the prolamellar body begins to reorganize into stacks of thylakoids, and chlorophyll starts to be produced. This process, where the etioplast becomes a chloroplast, takes several hours.Gymnosperms do not require light to form chloroplasts.

Light, however, does not guarantee that a proplastid will develop into a chloroplast. Whether a proplastid develops into a chloroplast some other kind of plastid is mostly controlled by the nucleus and is largely influenced by the kind of cell it resides in.

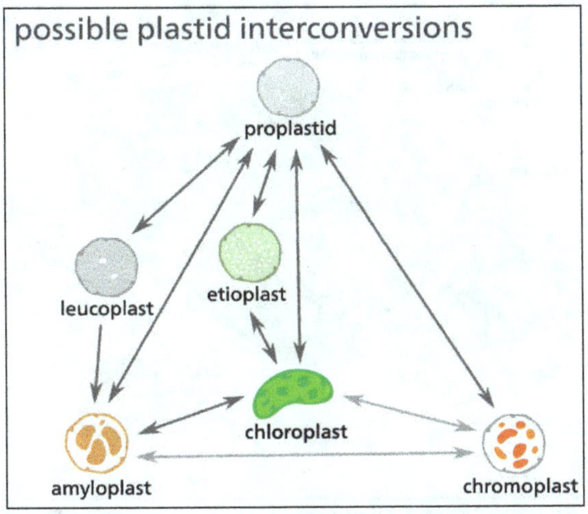

Many plastid interconversions are possible

Plastid Interconversion

Plastid differentiation is not permanent, in fact many interconversions are possible. Chloroplasts may be converted to chromoplasts, which are pigment-filled plastids responsible for the bright colors seen in flowers and ripe fruit. Starch storing amyloplasts can also be converted to chromoplasts, and it is possible for proplastids to develop straight into chromoplasts. Chromoplasts and amyloplasts can also become chloroplasts, like what happens when a carrot or a potato is illuminated. If a plant is injured, or something else causes a plant cell to revert to a meristematic state, chloroplasts and other plastids can turn back into proplastids. Chloroplast, amyloplast, chromoplast, proplast, etc., are not absolute states—intermediate forms are common.

Chloroplast Division

Most chloroplasts in a photosynthetic cell do not develop directly from proplastids or etioplasts. In fact, a typical shoot meristematic plant cell contains only 7–20 proplastids. These proplastids differentiate into chloroplasts, which divide to create the 30–70 chloroplasts found in a mature photosynthetic plant cell. If the cell divides, chloroplast division provides the additional chloroplasts to partition between the two daughter cells.

In single-celled algae, chloroplast division is the only way new chloroplasts are formed. There is no proplastid differentiation—when an algal cell divides, its chloroplast divides along with it, and each daughter cell receives a mature chloroplast.

Almost all chloroplasts in a cell divide, rather than a small group of rapidly dividing chloroplasts. Chloroplasts have no definite S-phase—their DNA replication is not synchronized or limited to that of their host cells. Much of what we know about chloroplast division comes from studying organisms like *Arabidopsis* and the red alga *Cyanidioschyzon merolæ*.

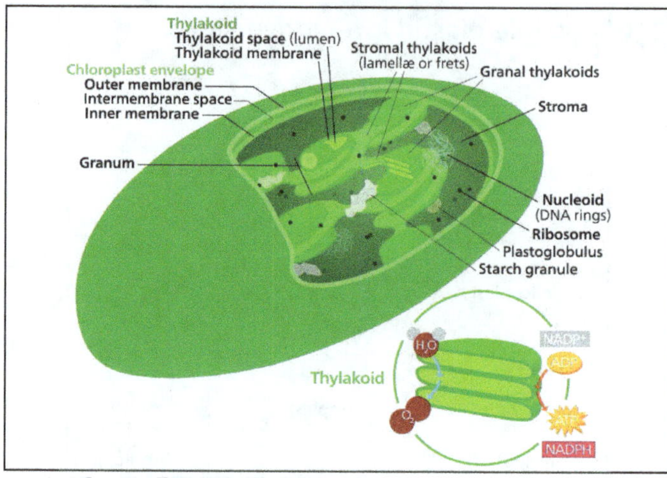

Most chloroplasts in plant cells, and all chloroplasts in algae arise from chloroplast division

The division process starts when the proteins FtsZ1 and FtsZ2 assemble into filaments, and with the help of a protein ARC6, form a structure called a Z-ring within the chloroplast's stroma. The Min system manages the placement of the Z-ring, ensuring that the chloroplast is cleaved more or less evenly. The protein MinD prevents FtsZ from linking up and forming filaments. Another protein ARC3 may also be involved, but it is not very well understood. These proteins are active at the poles of the chloroplast, preventing Z-ring formation there, but near the center of the chloroplast, MinE inhibits them, allowing the Z-ring to form.

Next, the two plastid-dividing rings, or PD rings form. The inner plastid-dividing ring is located in the inner side of the chloroplast's inner membrane, and is formed first.The outer plastid-dividing ring is found wrapped around the outer chloroplast membrane. It consists of filaments about 5 nanometers across, arranged in rows 6.4 nanometers apart, and shrinks to squeeze the chloroplast. This is when chloroplast constriction begins. In a few species like *Cyanidioschyzon merolæ*, chloroplasts have a third plastid-dividing ring located in the chloroplast's intermembrane space.

Late into the constriction phase, dynamin proteins assemble around the outer plastid-dividing ring, helping provide force to squeeze the chloroplast.Meanwhile, the Z-ring and the inner plastid-dividing ring break down.During this stage, the many chloroplast DNA plasmids floating around in the stroma are partitioned and distributed to the two forming daughter chloroplasts.

Later, the dynamins migrate under the outer plastid dividing ring, into direct contact with the chloroplast's outer membrane, to cleave the chloroplast in two daughter chloroplasts.

A remnant of the outer plastid dividing ring remains floating between the two daughter chloroplasts, and a remnant of the dynamin ring remains attached to one of the daughter chloroplasts.

Of the five or six rings involved in chloroplast division, only the outer plastid-dividing ring is present for the entire constriction and division phase—while the Z-ring forms first, constriction does not begin until the outer plastid-dividing ring forms.

Regulation

In species of algae that contain a single chloroplast, regulation of chloroplast division is extremely important to ensure that each daughter cell receives a chloroplast—chloroplasts can't be made from scratch. In organisms like plants, whose cells contain multiple chloroplasts, coordination is looser and less important. It is likely that chloroplast and cell division are somewhat synchronized, though the mechanisms for it are mostly unknown.

Light has been shown to be a requirement for chloroplast division. Chloroplasts can grow and progress through some of the constriction stages under poor quality green light, but are slow to complete division—they require exposure to bright white light to complete division. Spinach leaves grown under green light have been observed to contain many large dumbbell-shaped chloroplasts. Exposure to white light can stimulate these chloroplasts to divide and reduce the population of dumbbell-shaped chloroplasts.

Chloroplast Inheritance

Like mitochondria, chloroplasts are usually inherited from a single parent. Biparental chloroplast inheritance—where plastid genes are inherited from both parent plants—occurs in very low levels in some flowering plants.

Many mechanisms prevent biparental chloroplast DNA inheritance, including selective destruction of chloroplasts or their genes within the gamete or zygote, and chloroplasts from one parent being excluded from the embryo. Parental chloroplasts can be sorted so that only one type is present in each offspring.

Gymnosperms, such as pine trees, mostly pass on chloroplasts paternally, while flowering plants often inherit chloroplasts maternally. Flowering plants were once thought to only inherit chloroplasts maternally. However, there are now many documented cases of angiosperms inheriting chloroplasts paternally.

Angiosperms, which pass on chloroplasts maternally, have many ways to prevent paternal inheritance. Most of them produce sperm cells that do not contain any plastids. There are many other documented mechanisms that prevent paternal inheritance in these flowering plants, such as different rates of chloroplast replication within the embryo.

Among angiosperms, paternal chloroplast inheritance is observed more often in hybrids than in offspring from parents of the same species. This suggests that incompatible hybrid genes might interfere with the mechanisms that prevent paternal inheritance.

Transplastomic Plants

Recently, chloroplasts have caught attention by developers of genetically modified crops. Since, in most flowering plants, chloroplasts are not inherited from the male parent, transgenes in these plastids cannot be disseminated by pollen. This makes plastid transformation a valuable tool for the creation and cultivation of genetically modified plants that are biologically contained, thus posing significantly lower environmental risks. This biological containment strategy is therefore suitable for establishing the coexistence of conventional and organic agriculture. While the reliability of this mechanism has not yet been studied for all relevant crop species, recent results in tobacco plants are promising, showing a failed containment rate of transplastomic plants at 3 in 1,000,000.

Etioplast

Etioplasts are chloroplasts that have not been exposed to light. They are usually found in flowering plants (Angiosperms) grown in the dark. If a plant is kept out of light for several days, its normal chloroplasts will actually convert into etioplasts. Etioplasts lack active pigment and can technically be considered leucoplasts. High concentrations of etioplasts will cause leaves to appear yellow rather than green.

These plant organelles contain prolamellar bodies, which are membrane aggregations of semi-crystalline lattices of branched tubules that carry the precursor pigment for chlorophyll. The prolamellar bodies are often (and presumed always) arranged in geometric patterns.

They are converted to chloroplasts via the stimulation of chlorophyll synthesis by the plant hormone cytokinin soon after exposure to light. Thylakoids and grana arise from the prolamellar bodies during this process.

Gerontoplast

A gerontoplast is a plastid found in formerly green tissues that are currently senescing. A gerontoplast is a chloroplast that has re-purposed through a process of developmental senescence.

Transformation of Chloroplasts to Gerontoplasts

The term *gerontoplast* was first introduced by Sitte (1977) to define the unique features of the plastid formed during leaf senescence. The process of senescence brings about regulated dismantling of cellular organelles. Chloroplast shows the first sign of senescence induced degradation and is the last organelle to survive when other organelles are completely disorganized. The formation of gerontoplast from chloroplast during senescence involves extensive structural modifications of thylakoid membrane with the concomitant formation of a large number of plastoglobuli with lipophilic materials. The envelope of the plastid, however, remains intact.

Chromoplasts

The coloration of the petals and sepals on the Bee orchid is controlled by a specialized organelle in plant cells called a chromoplast.

Chromoplasts are plastids, heterogeneousorganelles responsible for pigmentsynthesis and storage in specific photosynthetic eukaryotes. It is thought that like all other plastids including chloroplasts and leucoplasts they are descended from symbioticprokaryotes.

Function

Chromoplasts are found in fruits, flowers, roots, and stressed and aging leaves, and are responsible for their distinctive colors. This is always associated with a massive increase in the accumulation of carotenoid pigments. The conversion of chloroplasts to chromoplasts in ripening is a classic example.

They are generally found in mature tissues and are derived from preexisting mature plastids. Fruits and flowers are the most common structures for the biosynthesis of carotenoids, although other reactions occur there as well including the synthesis of sugars, starches, lipids, aromatic compounds, vitamins and hormones. The DNA in chloroplasts and chromoplasts is identical. One subtle difference in DNA was found after a liquid chromatography analysis of tomato chromoplasts was conducted, revealing increased cytosine methylation.

Chromoplasts synthesize and store pigments such as orange carotene, yellow xanthophylls, and various other red pigments. As such, their color varies depending on what pigment they contain. The main evolutionary purpose of chromoplasts is probably to attract pollinators or eaters of colored fruits, which help disperse seeds. However, they are also found in roots such as carrots and sweet potatoes. They allow the accumulation of large quantities of water-insoluble compounds in otherwise watery parts of plants.

When leaves change color in the autumn, it is due to the loss of green chlorophyll, which unmasks preexisting carotenoids. In this case, relatively little new carotenoid is produced—the change in plastid pigments associated with leaf senescence is somewhat different from the active conversion to chromoplasts observed in fruit and flowers.

There are some species of flowering plants that contain little to no carotenoids. In such cases there are plastids present within the petals that closely resemble chromoplasts and are sometimes visually indistinguishable. Anthocyanins and flavonoids located in the cell vacuoles are responsible for other colors of pigment.

The term "chromoplast" is occasionally used to include *any* plastid that has pigment, mostly to emphasize the difference between them and the various types of leucoplasts, plastids that have no pigments. In this sense, chloroplasts are a specific type of chromoplast. Still, "chromoplast" is more often used to denote plastids with pigments other than chlorophyll.

Structure and Classification

Using a light microscope chromoplasts can be differentiated and are classified into four main types. The first type is composed of proteic stroma with granules. The second is composed of protein crystals and amorphous pigment granules. The third type is composed of protein and pigment crystals. The fourth type is a chromoplast which only contains crystals. An electron microscope reveals even more, allowing for the identification of substructures such as globules, crystals, membranes, fibrils and tubules. The substructures found in chromoplasts are not found in the mature plastid that it divided from.

The presence, frequency and identification of substructures using an electron microscope has led to further classification, dividing chromoplasts into five main categories: Globular chromoplasts, crystalline chromoplasts, fibrillar chromoplasts, tubular chromoplasts and membranous chromoplasts. It has also been found that different types of chromoplasts can coexist in the same organ. Some examples of plants in the various categories include mangos, which have globular chromoplasts, and carrots which have crystalline chromoplasts.

Although some chromoplasts are easily categorized, others have characteristics from multiple categories that make them hard to place. Tomatoes accumulate carotenoids, mainly lycopene crystalloids in membrane-shaped structures, which could place them in either the crystalline or membranous category.

Evolution

Plastids are descendants of cyanobacteria, photosynthetic prokaryotes, which integrated themselves into the eukaryotic ancestor of algæ and plants, forming an endosymbiotic relationship. The ancestors of plastids diversified into a variety of plastid types, including chromoplasts. Plastids also possess their own small genome and some have the ability to produce a percentage of their own proteins.

The main evolutionary purpose of chromoplasts is to attract animals and insects to pollinate their flowers and disperse their seeds. The bright colors often produced by chromoplasts is one of many ways to achieve this. Many plants have evolved symbiotic relationships with a single pollinator. Color can be a very important factor in determining which pollinators visit a flower, as specific colors attract specific pollinators. White flowers tend to attract beetles, bees are most often attracted to violet and blue flowers, and butterflies are often attracted to warmer colors like yellows and oranges.

Research

Chromoplasts are not widely studied and are rarely the main focus of scientific research. They often play a role in research on the tomato plant (*Solanum lycopersicum*). Lycopene is responsible for the red color of a ripe fruit in the cultivated tomato, while the yellow color of the flowers is due to xanthophyllsviolaxanthin and neoxanthin.

Carotenoid biosynthesis occurs in both chromoplasts and chloroplasts. In the chromoplasts of tomato flowers, carotenoid synthesis is regulated by the genes Psyl, Pds, Lcy-b, and Cyc-b. These genes, in addition to others, are responsible for the formation of carotenoids in organs and structures. For example, the Lcy-e gene is highly expressed in leaves, which results in the production of the carotenoid lutein.

White flowers are caused by a recessive allele in tomato plants. They are less desirable in cultivated crops because they have a lower pollination rate. In one study, it was found that chromoplasts are still present in white flowers. The lack of yellow pigment in their petals and anthers is due to a mutation in the CrtR-b2 gene which disrupts the carotenoid biosynthesis pathway.

The entire process of chromoplast formation is not yet completely understood on the molecular level. However, electron microscopy has revealed part of the transformation from chloroplast to chromoplast. The transformation starts with remodeling of the internal membrane system with the lysis of the intergranal thylakoids and the grana. New membrane systems form in organized membrane complexes called thylakoid plexus. The new membranes are the site of the formation of carotenoid crystals. These newly synthesized membranes do not come from the thylakoids, but rather from vesicles generated from the inner membrane of the plastid. The most obvious biochemical change would be the downregulation of photosynthetic gene expression which results in the loss of chlorophyll and stops photosynthetic activity.

In oranges, the synthesis of carotenoids and the disappearance of chlorophyll causes the color of the fruit to change from green to yellow. The orange color is often added artificially—light yellow-orange is the natural color created by the actual chromoplasts.

Valencia oranges *Citris sinensis L* are a cultivated orange grown extensively in the state of Florida. In the winter, Valencia oranges reach their optimum orange-rind color while reverting to a green color in the spring and summer. While it was originally thought that chromoplasts were the final stage of plastid development, in 1966 it was proved that chromoplasts can revert to chloroplasts, which causes the oranges to turn back to green.

Compare

- Plastid
  - Chloroplast and etioplast
  - Chromoplast
  - Leucoplast
    - Amyloplast
    - Elaioplast
    - Proteinoplast

Leucoplast

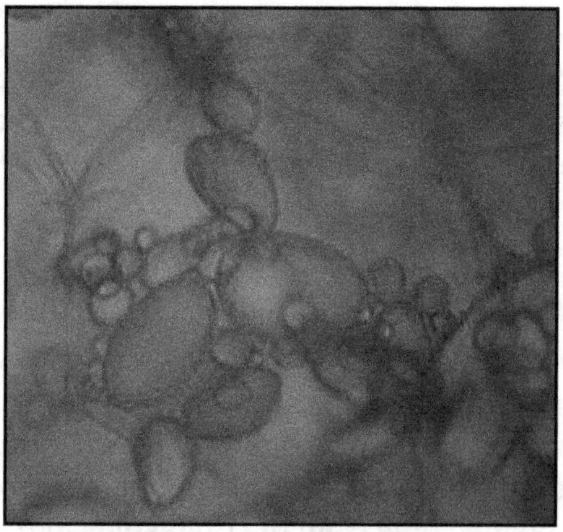

Leucoplasts, specifically, amyloplasts

Leucoplasts are a category of plastid and as such are organelles found in plant cells. They are non-pigmented, in contrast to other plastids such as the chloroplast.

Lacking photosynthetic pigments, leucoplasts are not green and are located in non-photosynthetic tissues of plants, such as roots, bulbs and seeds. They may be specialized for bulk storage of starch, lipid or protein and are then known as amyloplasts, elaioplasts, or proteinoplasts (also called aleuroplasts) respectively. However, in many cell types, leucoplasts do not have a major storage function and are present to provide a wide range of essential biosynthetic functions, including the synthesis of fatty acids such as palmitic acid, many amino acids, and tetrapyrrole compounds such as heme. In general, leucoplasts are much smaller than chloroplasts and have a variable morphology, often described as amoeboid. Extensive networks of stromules interconnecting leucoplasts have been observed in epidermal cells of roots, hypocotyls, and petals, and in callus and suspension culture cells of tobacco. In some cell types at certain stages of development, leucoplasts are clustered around the nucleus with stromules extending to the cell periphery, as observed for proplastids in the root meristem.

Etioplasts, which are pre-granal, immature chloroplasts but can also be chloroplasts that have been deprived of light, lack active pigment and can be considered leucoplasts. After several minutes exposure to light, etioplasts begin to transform into functioning chloroplasts and cease being leucoplasts. Amyloplasts are of large size and store starch. Proteinoplasts store proteins and are found in seeds (pulses). Elaioplasts store fats and oils and are found in seeds. They are also called oleosomes. [castor, groundnut] Etioplasts are plastids without pigments and store food and lamellar structures. These plastids occur in etiolated plants due to the absence of light.

Elaioplast

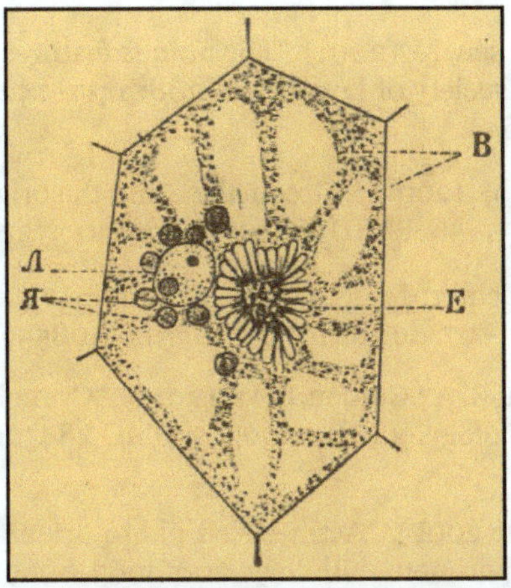

Illustration from Collegiate Dictionary, FA Brockhaus and IA Efron, circa 1905. Cell of very young leaf of *Vanilla planifolia*; E - elaioplasts; Л - the nucleus; Я - leucoplasts; B - vacuoles

Elaioplasts are a type of leucoplast that is specialized for the storage of lipids in plants. Elaioplasts house oil body deposits as rounded plastoglobuli, which are essentially fat droplets.

Being a variety of leucoplast, elaioplasts are non-pigmented and fall into the much broader organelle category of plantplastids. A different example of a storage-specialized leucoplast is the amyloplast, which stores starch.

Proteinoplast

Proteinoplasts (sometimes called *proteoplasts, aleuroplasts,* and *aleuronaplasts*) are specialized organelles found only in plant cells. Proteinoplasts belong to a broad category of organelles known as plastids. Because they lack pigment, proteinoplasts are more specifically a kind of leucoplast. They contain crystalline bodies of protein and can be the sites of enzyme activity involving those proteins. Proteinoplasts are found in many seeds, such as brazil nuts, peanuts and pulses. Although all plastids contain high concentrations of protein, proteinoplasts were identified in the 1960s and 1970s as having large protein inclusions that are visible with both light microscopes and electron microscopes.

A book written in 2007 noted that no scientific research had been published in the previous 25 years on proteinoplasts.

References

- Zhang, Q.; Sodmergen (2010). "Why does biparental plastid inheritance revive in angiosperms?". Journal of Plant Research. 123 (2): 201–206. PMID 20052516. doi:10.1007/s10265-009-0291-z
- Milo, Ron; Philips, Rob. "Cell Biology by the Numbers: How large are chloroplasts?". book.bionumbers.org. Retrieved 7 February 2017
- Viola, R.; Nyvall, P.; Pedersén, M. (2001). "The unique features of starch metabolism in red algae". Proceedings of the Royal society of London, B. 268: 1417–1422. PMC 1088757. PMID 11429143. doi:10.1098/rspb.2001.1644
- Lewis, L. A.; McCourt, R. M. (2004). "Green algae and the origin of land plants". American Journal of Botany. 91 (10): 1535–56. PMID 21652308. doi:10.3732/ajb.91.10.1535
- Schnepf, Eberhard; Elbrächter, Malte (1999). "Dinophyte chloroplasts and phylogeny – A review". Grana. 38 (2–3): 81–97. doi:10.1080/00173139908559217
- Kumar, K.; Mella-Herrera, R. A.; Golden, J. W. (2010). "Cyanobacterial Heterocysts". Cold Spring Harbor Perspectives in Biology. 2 (4): a000315. PMC 2845205. PMID 20452939. doi:10.1101/cshperspect.a000315
- Takagi, Shingo (December 2002). "Actin-based photo-orientation movement of chloroplasts in plant cells". Journal of Experimental Biology. 206: 1963–1969. doi:10.1242/jeb.00215. Retrieved 6 March 2013
- Berg JM; Tymoczko JL; Stryer L. (2002). Biochemistry. (5th ed.). W H Freeman. pp. Section 19.4. Retrieved 30 October 2012
- Keeling, Patrick J. (2004). "Diversity and evolutionary history of plastids and their hosts". American Journal of Botany. 91 (10): 1481–93. PMID 21652304. doi:10.3732/ajb.91.10.1481

- Steer, Brian E.S. Gunning, Martin W. (1996). Plant cell biology : structure and function. Boston, Mass.: Jones and Bartlett Publishers. p. 24. ISBN 0-86720-504-0
- J D, Rochaix (1998). The molecular biology of chloroplasts and mitochondria in Chlamydomonas. Dordrecht [u.a.]: Kluwer Acad. Publ. pp. 550–565. ISBN 978-0-7923-5174-0
- May, T. (2000). "14-3-3 Proteins Form a Guidance Complex with Chloroplast Precursor Proteins in Plants". The Plant Cell Online. 12: 53–64. doi:10.1105/tpc.12.1.53

Chapter 6

Signal Transduction in Plants

The process of transmitting a chemical or physical signal through a series of molecular events which result in a cellular response is termed as signal transduction. The various facets related to signal transduction in plants such as plant receptors, signaling pathways and secondary messengers are discussed in this chapter in an easy-to-understand manner.

Plant cells communicate to coordinate their activities in response to the changing conditions of light and temperature that guide the plant's cycle of growth, flowering, and fruiting. Plant cells also communicate to coordinate what goes on in their roots, stems, and leaves.

The cascade of signal on a cell surface that gets converted into a specific cellular response is called a signal transduction pathway. The process involves three stages: reception, transduction and response. In reception, a chemical signal binds to a cellular protein at the cell surface or inside the cell. In transduction, binding leads to a change in the receptor that triggers a series of changes in various relay molecules. In response, a specific cellular activity is triggered.

Successful adaptation to changing ecological conditions is a common challenge faces by all organisms. Multicellular life forms has to face one more level of challenge that is to develop egg or seed into a mature organisms which requires spatially and temporally regulated cell-cell signaling.

Though mutational approach has been a powerful technique for identification of genes encoding components of signal transduction pathways, classical biochemical approaches along with latest technological advances being made in proteomics and genomics are going to have a radical impact upon our understanding of how plants perceive and respond to signals.

It is natural; if multicellularity has evolved independently in plants and animals then the mechanisms for cell communication will also have evolved separately and would be expected to be different. Arabidopsis thaliana, a model system has completely sequenced genome, suggests that there are no homologs of Wnt, Notch, Jak/STAT, TGF-beta, Ras, or the nuclear receptor family in plants. Also cAMP has not been implicated in intracellular signaling. There is some resemblance between plant and animal signaling in the form of Nitric oxide and Ca^{2+}.

Most cell surface receptors in animals are G-protein-linked which are very rare in plants. Plants prefer mostly enzyme-linked receptors which are Tran's membrane receptor serine/theronine kinases as against receptor tyrosine kinases in animals.

Various plant hormones/growth regulators help to coordinate plant development. These are small molecules which diffuse readily through cell walls and may act locally or away. Most common examples include ethylene, auxin, gibberellins and abscisic acid. Each growth regulator may have multiple effects. The specific effect depends on which other growth regulators are acting, on environmental conditions, on the nutritional state of the plant, and on the responsiveness of the target cell.

In plants, there are various light responsive proteins to sense light of different wavelengths which are referred as photoreceptors/photo protein. All photo proteins sense light by means of a covalently attached light-absorbing chromosphere, which changes its shape in response to light and then induces a change in the protein's conformation. Signal transduction pathway from light absorption by photoreceptors to the light-dependent developmental changes that prepare a plant for photosynthetic activity/photo morphogenesis.

Plant Receptors

Unlike animal systems, where organs and glands are discrete, plant development is a continuum of growth and differentiation in which cells and tissues both produce and respond to many hormones. Biochemical analysis is further compromised by the cell wall and the plethora of secondary chemical products that plants synthesize. Fortunately, with the arrival of modern *Arabidopsis* genetics, mutations were identified that increased or decreased the response of a whole plant to a particular hormone. With the mutants in hand, genes could be isolated without the need for sophisticated biochemistry, and in the past few years, receptors to many of the major plant hormones have been identified. This has allowed biochemical studies and has also initiated a comprehensive comparison of how multicellular plants and animals, which most likely evolved independently, use similar organic-based hormones to organize development.

Two-component-based Hormone Receptors

Two-component regulators derive their name from the transfer of phosphates between a collection of signature proteins. This starts when the first component perceives an input signal to cause an autophosphorylation of a conserved histidine in the transmitter domain. The signal is then relayed through a phosphotransfer to a conserved aspartate residue in the receiver domain of the second component, often called the response regulator. Two-component regulators were first defined in prokaryotes, where they are quite prevalent, often representing up to 1% of the genome. To date, no two-component regulators have been found in animals; however, they have been identified in both fungi and plants. The fungal and plant two-component systems are more complicated than the canonical prokaryotic system in that the input histidine kinase domain itself often carries a receiver domain. Moreover, an intermediate, designated HPt for histidine phosphotransfer protein can also participate in the phosphate transfer between the histidine kinase and the receiver domain. Three-dimensional protein studies show that HPt domains share similarity with parts of the transmitter domain.

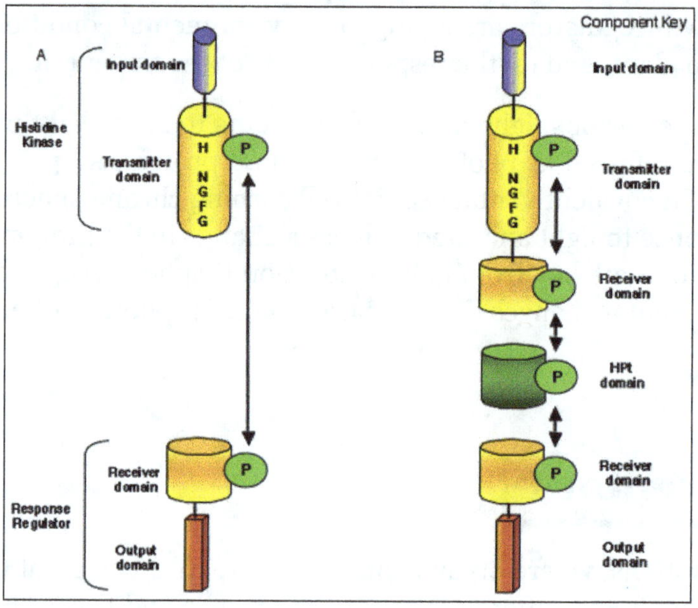

The prototypical two-component system (*A*) The two-step system found only in prokaryotes. The input domain binds a signal molecule, resulting in the phosphorylation (green-encircled P) of a conserved histidine in the transmitter domain. This phosphoryl group is subsequently transferred to the receiver domain on the response regulator. As a result, the output domain is activated to transduce the signal. (*B*) The multistep derivation of the two-component system found in prokaryotes and some eukaryotes. Overall phosphotransfer is similar to *A* with the exception of intermediate components acting between the transmitter and output domains. These include receiver domains fused to the transmitter domain and a separate intermediate Hpt domain that can receive a phosphoryl group from a receiver domain. In all cases, phosphotransfers can be bidirectional. H (histidine), N (asparagine), G (glycine), and F (phenylalanine) represent conserved blocks characteristic of functional histidine kinases.

In Arabidopsis, eight two-component input regulators genes have been functionally identified as hormone receptors. Three are involved in cytokinin perception, and five encode ethylene receptors. The first cytokinin receptor was identified by taking advantage of the ability of exogenous cytokinins to cause greening in *Arabidopsis* plant tissue culture. An insensitive mutant was isolated, which led to the identification of the membrane-bound two-component input regulator, CRE1. Subsequently, two close homologs, *AHK2* and *AHK3*, were also identified. All three gene products have been shown to bind cytokinin and complement yeast and bacterial two-component mutants in a cytokinin-dependent manner. In *Arabidopsis,* there are five genes that encode HPt proteins, designated AHPs, and it appears that these intermediates transmit the membrane perceived cytokinin signal to the response regulators (ARRs) located in the nucleus.

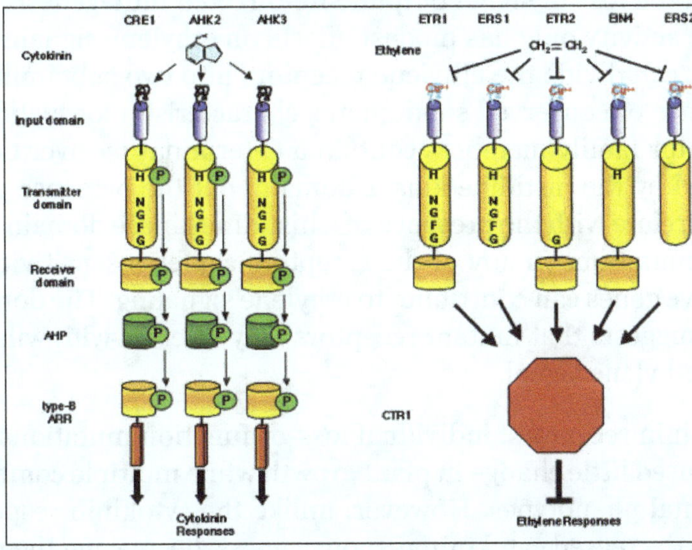

Structural relationships within and between two-component regulators involved in cytokinin and ethylene perception (*A*) In *Arabidopsis,* three cytokinin receptors (CRE1, AHR2, and AHR3) all bind cytokinins at their extracellular input domain. This results in a phospho-relay that eventually activates cytokinin responses. AHP and the type-B ARR are the *Arabidopsis* equivalents of the Hpt domain and response regulator proteins, respectively. (*B*) Five ethylene receptors (ETR1, ERS1, ETR2, EIN4, and ERS2) exist in *Arabidopsis* that all bind ethylene in their extracellular domain. Various forms of the transmitter domain exist on each receptor so, in principle, not all receptors can transfer phosphate groups. However, genetic analysis suggests that all of the ethylene receptors activate their downstream target component (CTR1) in the absence of the gas, and that binding of ethylene to the receptors inhibits this activation of CTR1.

Because a small gene family defines the cytokinin receptors, it was perhaps not surprising that single loss-of-function mutations were not lethal and, for the most part, allowed plants to grow normally. Combinations of double and triple mutations did expose abnormal developmental phenotypes, but surprisingly, embryogenesis and early seedling patterning were relatively normal. Although it is possible that other cytokinin perception systems exist, these results hint that the traditional view of an essential role for cytokinins in plant growth and development may be exaggerated.

The first genes encoding ethylene receptors were also identified through genetic screens for mutants that did not respond to hormone application. These studies led to the classification of five genes (*ETR1, ETR2, ERS1, ERS2, EIN4*) that share sequence similarity with those enocoding the two-component input regulators. Heterologous expression studies in yeast have clearly defined the signal input domain that perceives ethylene as a hydrophobic region at the N terminus of members of the ethylene receptor family. However, unlike the cytokinin story, the transmitter domain has been more difficult to understand for a variety of reasons. First, although truncation of the histidine kinase domain results in the loss of signal output, site-directed mutagenesis that specifically eliminates kinase activity only has modest effects on ethylene signaling. Second, phylogenetic analysis can divide the ethylene receptors into two subfamilies based on the presence or absence of conserved subdomains characteristic for histidine kinases. Finally, some receptor family members contain a C-terminal receiver domain that may be phosphorylated by the histidine kinase domain, but the presence of a receiver domain does not correlate with the presence of a histidine kinase domain. However, since gain-of-function mutations in any of the receptors confer insensitivity to ethylene, it appears that all five genes can contribute to ethylene signaling. The dominant nature of these mutations suggests that mutant receptors may interact with wild-type receptors to transduce the ethylene signal.

As with the cytokinin receptors, individual loss-of-function mutations in the ethylene receptor genes caused little change in plant growth while multiple combinations resulted in developmental phenotypes. However, unlike the cytokinin response where loss of more receptors decreased cytokinin responsiveness, decreasing the number of functional ethylene receptors resulted in plants with a constitutive ethylene response. The most parsimonious explanation for this result is that the family of ethylene receptors normally functions in the absence of ethylene and that addition of the gas inhibits this function. This model fits well with the immediate downstream component of the receptors, CTR1, since this serine/threonine protein kinase is a negative regulator of the ethylene response in *Arabidopsis*. In the absence of ethylene, the receptors activate CTR1, which in turn suppresses the ethylene response. When ethylene binds the receptors, CTR1 activation is inhibited, releasing downstream components to activate the ethylene pathway. Protein interaction studies and colocalization of both ETR1 and CTR1 protein to the endoplasmic reticulum (ER) support the genetic interactions. Because specific mutations in *CTR1* that disrupt ETR1–CTR1 interactions cause movement of CTR1 to the cytosol and genetic removal of two or more ethylene receptors also results in CTR1 protein redistribution, it appears that the ethylene receptors act as a scaffolding protein for CTR1 at the ER. Therefore, ethylene receptors are quite unlike cytokinin or other two-component receptors in that there is no phosphor-relay to a typical HPt domain protein or a response regulator.

The oddities of the ethylene "two-component" receptors also apply to the way ethylene itself interacts with the receptor. Like other hormone ligands, ethylene is active at nanomolar concentrations. However, unlike most hormones, which are generally complex with respect to molecular shape, ethylene simply consists of two carbons. The lack of multiple contact points between ethylene and its receptor would not allow for the many weak interactions that are usually required to achieve high-affinity ligand–receptor specificity. This peculiarity was explained by the demonstration that a copper ion associates with the ethylene receptor and is required for binding of ethylene in both *Arabidopsis* and heterologous yeast systems expressing *ETR1*. Interestingly, the cyanobacterium *Synechocystis* strain 6803 contains an ethylene receptor ortholog, which also requires a copper cofactor to bind ethylene at a high affinity. Although *Synechocystis* does not synthesize or respond to ethylene, the conservation between bacterial and higher plant genes suggests that the progenitor to the ethylene receptor in plants was co-opted from a cyanobacterial ancestor after plant cells acquired the chloroplast during endosymbiosis.

This model implies that the progenitor ethylene receptor may have evolved to influence an already existing plant signaling pathway. Models of signaling pathway evolution have suggested that these pathways evolve in reverse order from the target transcription factor to the first step, which is the receptor and ligand combination. Often new components act as neomorphic and dominant-negative units that reverse the function of the previous step. This would appear to hold true for the upper part of the ethylene signaling pathway in which CTR1 acts negatively on EIN2, ethylene receptors act positively on CTR1, and, finally, ethylene acts negatively on the receptors. The idea that the ethylene receptor was a late addition to a pre-existing pathway may explain the major paradox in ethylene signaling regarding how enhanced sensitivity can occur with a reduction in receptor number. The observation that glucose can enhance the degradation of EIN3, a key transcriptional regulator at the bottom of the ethylene response pathway, hints that the ethylene pathway has other modulators. If the role of the bottom of the ethylene response pathway is to modulate sugar responses, which in turn determines cell growth and expansion, then the role of ethylene and its upstream receptor may simply be to inhibit a carbon response pathway.

Leucine-rich Repeat (LRR)-based Hormone Receptors

In both plants and animals, steroid-based molecules have been identified that act as hormones for directing growth and development. In plants, the best described steroid hormone is brassinosteroid (BR). The first interest in BR arose from its growth promoting qualities upon application. The isolation of *Arabidopsis* mutants deficient in BR biosynthesis clearly demonstrated that BR is in fact a general regulator of plant growth and development. The identification of mutations affecting BR perception involved screening for *Arabidopsis* mutants that showed normal root growth on exogenous BR or by screening for dwarf mutants that mimicked BR auxotrophic phenotypes that were not rescued by exogenous BR.

In both cases, these approaches uncovered many alleles of a single locus, designated *BRI1*, which encodes a plasma membrane-associated LRR receptor-like kinase (LRR-RLK). The LRR is a common protein motif of usually 20–29 residues that is found in a number of proteins with diverse functions throughout all kingdoms. Plant receptors have an extracellular LRR domain that is fused to an intracellular serine/threonine kinase domain. In animals, LRR-based receptors transduce their signals through intracellular protein–protein interactions rather than directly through kinase domains. For example, the activated Toll receptor in *Drosophila* recruits, among other proteins, a serine/threonine kinase that shares similarity to the kinase domain found in plant LRR-RLKs. Interestingly, other examples exist where plants have fused regulatory domains into a single protein while the equivalent domains in animals must associate through protein–protein interactions.

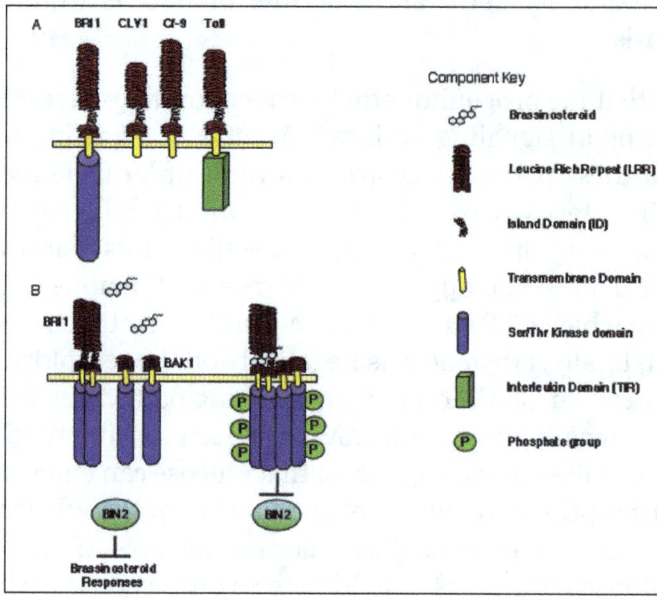

LRR receptors (*A*) Comparison of four types of LRR receptors found in plants and animals: BRI1, an *Arabidopsis* BR receptor; CLV1, a receptor involved in binding the small peptide CLV3 that down-regulates meristem size in *Arabidopsis*; Cf-9, a receptor involved in plant pathogen responses in tomato; and Toll, an embryo patterning and innate immunity receptor in *Drosophila*. (*B*) Mechanism of BR perception. Normally the BRI1 and BAK1 LRR kinases are inactive homodimers. Upon BR binding, the BRI1 and BAK1 proteins aggregate, resulting in an autophosphorylation event. This leads to the repression of the downstream negative regulator, BIN2, and induction of BR responses.

The uncovering of a LRR-RLK as having a role in plant hormone signaling is perhaps not unexpected since this class of genes constitutes a large family consisting of >200 genes in the *Arabidopsis* genome. What was unexpected is that the extracellular LRR module directly binds BR. It is commonly thought that the LRR motif functions as a protein–protein interaction domain rather than as a domain that interacts with organic molecules. In the case of the BRI1 LRR, however, the 25 leucine repeats are interrupted by a unique 70-amino-acid island domain (ID) between repeat 21 and 22. The importance of the ID in BR signaling was implied by the observation that mutations that confer BR insensitivity cluster to this region and, more directly, by the ability of BR to bind to the ID in vitro. Perhaps the ID explains the unique property of this extracellular domain to bind organic molecules, and it has been speculated that LRR22-ID may not fold in a typical way.

In animals, a superfamily of nuclear receptors exists that are activated when the fully buried ligand completes the hydrophobic core of the receptor. In this scenario, the ligand acts to stabilize the protein so that it can become nuclear localized and directly bind DNA. In plants, it appears that binding of BR to inactive BRI1 homodimers induces autophosphorylation of its cytoplasmic kinase domain and thereby stimulates the interaction of BRI1 with another related LRR-RLK called BAK1. BAK1 contains a short four-leucine zipper repeat extracellular domain with no ID motif so it appears this LRR-RLK does not bind BR. Two models have been proposed for BRI1/BAK1 interactions. The first suggests that BR allows BRI1 and BAK1 to interact, thereby allowing transphosphorylation of BAK1. A second model is that BRI1 and BAK1 form a complex that is stabilized and activated by BR. Recently, in vitro analysis suggested that heteromeric associations between BRI1 and BAK1, as well as phosphorylation, are dependent on BR. Whatever the case, it appears that plants perceive steroid hormones at the plasma membrane and eventually stimulate gene expression through an involved signal transduction pathway that is similar to Wnt signaling in animals.

Binding experiments with BR indicate that it directly interacts with and activates the BRI1/BAK1 complex; however, some evidence exists to suggest that the first stages of signaling may be more complex. First, overexpressing the carboxypeptidase BRS1, which normally localizes to the extracellular space, can suppress a weak *bri1* allele. This suppression, which is specific to BRI1 and requires both BR and a functional protein kinase domain, indicates that BRS1 is involved in an early event in BRI1-mediated signaling. The molecular identity of BRS1 suggests that this protein may be involved in receptor proteolytic processing, and there is evidence that BRI1 may undergo endocytosis after binding BR.

Alternatively, BRS1 may be involved in the processing of an as of yet unidentified protein ligand. The latter idea gains some support from studies in tomato showing that the BRI1 homolog is also the receptor for an 18-amino-acid protein called systemin. Systemin is the processed form of an inactive 200-amino-acid preprotein made during insect herbivory to activate a battery of defense genes. Tomato *bri1* mutants, which are phenotypically similar to BR-deficient dwarfs, are defective in systemin signaling. Furthermore, although BR does not reduce systemin binding, it does inhibit the systemin response. It is possible that BR inhibits systemin signaling by directing the tomato BRI1 receptor away from systemin signaling and toward hormone signaling. Alternatively, BR and systemin may function through the same signaling pathway. In *Arabidopsis,* one function of a BR-bound BRI1 receptor is to negatively regulate the GSK3/SHAGGY-like kinase, BIN2. In the absence of BR, BIN2 negatively regulates activators of BR gene expression. In tomato, the BRI1 receptor is activated by systemin, which could influence BIN2 function. In this case, mutations in BIN2 and other downstream components of BR signaling may have altered systemin signaling. Although systemin has only been identified in some members of the Solanceae family, this observation of a shared receptor for both a peptide and organic molecule could mean that BR receptors in other plants may also have shared perception functions. Multiple recognition specificity may explain the discrepancy between the level of BRI protein, which is abundant in all tissues, and the low levels of BR ligand.

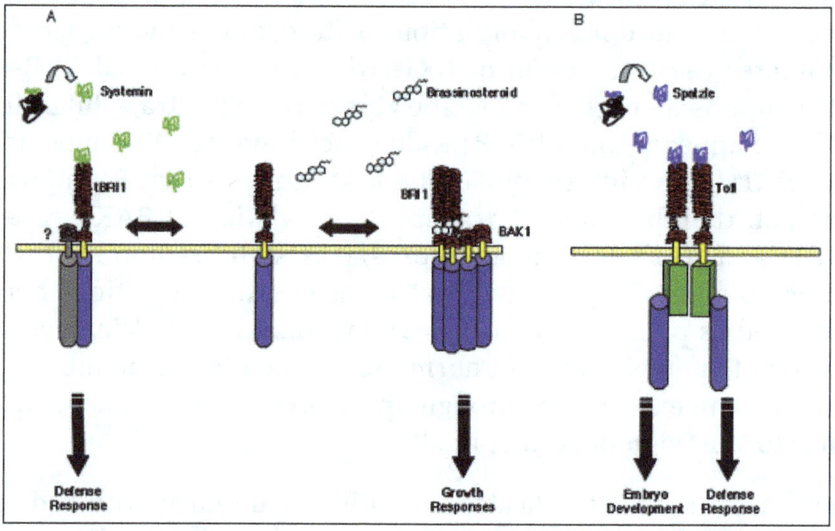

Consequences of LRR receptor–ligand interactions (*A*) In tomato, the BRI1 receptor (tBRI1) can bind either BR or the small peptide systemin. Systemin is processed from a larger polypeptide by an as of yet uncharacterized protease (curved arrow). The resulting ligand binds tBRI1, causing heterodimerization of the tBRI1 with an uncharacterized LRR protein (?) to induce defense responses. Alternatively, if tBRI1 binds BR, this results in a BRI–BAK1 dimerization and BR-dependent growth. (*B*) In *Drosophila*, a protease (curved arrow) cleaves a preprotein to generate a polypeptide Spatzle, which subsequently binds the Toll receptor. The choice between embryo development and innate immune response is determined by the developmentally regulated cleavage of the pre-Spatzle protein.

Aside from the structural similarities, other intriguing similarities exist between systemin/BR perception and signaling in tomato and Toll-based signaling in *Drosophila*. Like the tomato systemin/BR receptor, the Toll receptor also has two functions: one in embryogenesis, and the other in adult immune responses. Early events also appear to involve a proteolytic processing of an inactive preprotein to produce the activated Spatzle ligand that subsequently binds the extracellular LRR domain of the receptor. This leads to the recruitment of a serine/threonine kinase, which phosphorylates downstream components. In the case of the Toll pathway, however, a single ligand Spatzle is used in both processes but is cleaved by different developmentally regulated proteases into its mature form. The processing of the systemin ligand parallels that of the Toll ligand, as does the identification of a carboxyl peptidase in BRI1 signaling. Although not conserved at the amino acid level, the Toll receptor also contains an "island" within the LRR extracellular domains whose functions are unknown. Perhaps like BRI1, this region can also bind an organic molecule that modulates its function.

Ubiquitination-based Hormone Receptors

The first hints that some plant hormones had unique mechanisms of signaling came from the identification of genes required for normal responsiveness to exogenous auxin. Broadly speaking, these genes encoded two classes of molecules. One comprised members of a family of extremely shortlived nuclear proteins called Aux/IAAs. Aux/IAA genes are rapidly induced by auxin application, and their protein products are then quickly degraded. Mutations in Aux/IAA genes that conferred increased resistance to auxin mapped to a common domain shown to be important in protein stability. Hence, decreasing the turnover of an Aux/IAA protein somehow conferred decreased auxin sensitivity. The second class of molecules identified various components of the SCF (Skp1/Cullin/F-box) ubiquitin protein complex.

The SCF complex, which was first identified in yeast, provides a mechanism by which cells can rapidly degrade proteins by targeting them to the 26S proteosome. Briefly, ubiquitin is first activated by an E1 enzyme and then passed to a family of ubiquitin-conjugating E2 enzymes. The E2 enzyme carrying an ubiquitin molecule can interact with the Cullin subunit of the SCF complex. The SKP1 subunit acts as an adaptor protein between Cullin and the F-box so that when the F-box protein captures a protein target it is now in the vicinity of the E2 enzyme and can be ubiquinated. This process is reiterated until the target protein accumulates more than four ubiquitins, which signal it for degradation via the 26S proteosome.

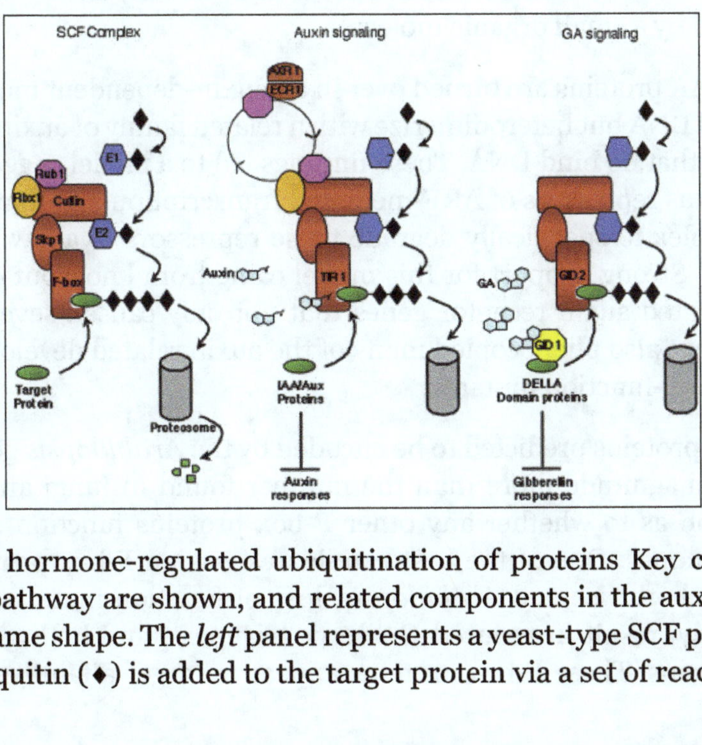

SCF-dependent hormone-regulated ubiquitination of proteins Key components in a canonical SCF pathway are shown, and related components in the auxin and GA pathways have the same shape. The *left* panel represents a yeast-type SCF proteolysis model in which an ubiquitin (♦) is added to the target protein via a set of reactions. Ubiquitin is activated by an E1 enzyme and conjugated to the target via the E2 and F-box proteins associated with the SCF complex. The F-box protein determines specificity of targets. As ubiquitination of the target is a reiterative process, the ubiquitin tail grows and, after reaching a certain length, causes the degradation of the target via the 20S proteosome. The *middle* panel depicts auxin-mediated degradation of Aux/IAA repressor targets. The TIR1 F-box protein binds auxin, and this stimulates its association with the Aux/ IAA protein. Targeting of Aux/IAA to the proteosome releases auxin response genes from repression. The modification of the Cullin subunit with the ubiquitin-like protein RUB1 and other proteins enhances SCF-dependent ubiquitination. The first auxin resistant mutant (*arx1*) identified a protein required in rubinylation. The *right* panel depicts GA-mediated degradation of the DELLA domain repressor targets. The GID1 protein binds the DELLA protein in a GA-dependent manner and targets it to the SCF^{GID2} complex for ubiquitination. Degradation of the DELLA domain proteins releases GA-dependent genes from repression.

There are other subunits and modifications that influence SCF function, but the specificity of the F-box protein in determining target substrate suggested that the SCF complex is central to the regulation of Aux/IAA protein levels. An F-box protein, TIR1, which specifically targets the Aux/IAA proteins for proteolysis, was identified in a screen for resistance to auxin transport inhibitors, and subsequent analysis suggested it was an essential component of auxin responsiveness. The SCF^{TIR1} complex was shown to bind Aux/IAA proteins directly, and this binding was enhanced in a cell free assay by the addition of auxin. Finally, it was shown that TIR1 itself could specifically bind bioactive auxin, and this resulted in an increased affinity for Aux/IAA proteins. Thus TIR1 is an auxin receptor, and unlike most F-box substrate interactions, this F-box protein can be directly modified by a small organic molecule.

Although Aux/IAA proteins are turned over in an auxin-dependent manner, these proteins do not bind DNA but heterodimerize with a related family of auxin response factor proteins (ARFs) that do bind DNA. These findings led to a model suggesting that Aux/ IAA proteins act as repressors of ARF-mediated transcription and that auxin activates the SCF^{TIR1} complex to specifically degrade these repressors to allow ARF-dependent auxin responses. Strong support for this model came from knockout combinations of all four TIR1-related auxin receptor genes that not only caused severe loss of auxin responsiveness but also phenocopied many of the auxin related developmental defects seen in ARF loss-of-function mutants.

The ~700 F-box proteins predicted to be encoded by the *Arabidopsis* genome is roughly one order of magnitude more than the number found in fungi and animals. This begs the question as to whether any other F-box proteins function as receptors for other plant hormones. One of the best candidates was an F-box protein involved in GA signaling. In both rice and *Arabidopsis,* loss-of-function mutations in *GID2* and *SLY1,* respectively, result in reduced GA responsiveness, and both genes encode related F-box proteins. These F-box proteins appear to target a family of proteins that normally function to repress GA-related processes. Because the addition of GA causes these proteins to be degraded and thereby activates GA-dependent gene expression, the GA system has many similarities to the components involved in auxin signaling. This led to suggestions that perhaps GID2/SLY1 F-boxes could be GA receptors. However, a loss-of-function mutation in a rice gene encoding a protein with similarity to animal hormone lipases (*GID1*) was also identified, and this mutation caused both a reduction in GA responsiveness and prevented the degradation of the DELLA domain repressor protein SLR1. More importantly, GID1 specifically bound bioactive GA, and its ability to bind SLR1 in a yeast two-hybrid assay was GA dependent. Hence, the addition of GA results in a protein–protein interaction between GID1 and the SLR1 DELLA domain protein to cause the degradation of this negative regulator of the GA response. The GID1 protein has no detectable lipase activity but is soluble and localizes to the nucleus where DELLA domain proteins reside. Interestingly, members of the TIR1 family of proteins also show nuclear localization, which raises questions of where these receptors encounter their hormones and whether the hormone influences receptor localization.

The identification of the SCF complexes and protein turnover as central to early auxin and GA signaling events suggests that the proteolysis of repressor proteins is an important hormone signal transduction mechanism in plants. However, unlike most epigenetic signaling events, such as a phosphotransfer, proteolysis is irreversible and therefore enforces directionality on a system. Such directionality is important in decisions that require total commitments such as controlling the cell cycle. By analogy, it could be argued that changes in levels of hormones such as auxin and GA during plant development should be followed by irreversible developmental commitments to processes such as cell division and expansion.

With respect to plant hormone signaling, the auxin and GA receptors are unique in that the number of components between the hormone ligand and the transcriptional factors they control is very small. Depending on the process being regulated, the length of the signaling pathway should reflect the required balance between the speed and robustness of a simple pathway versus the noise buffering ability of a more complex pathway. Buffering occurs because the addition of signaling components create opportunities for loops to form that can enhance robustness while reducing signaling noise. Hence, auxin and GA signaling should be very fast but with less buffering capacity than ethylene, BR, and cytokinin signaling. This modeling information should be useful when mapped onto the different physiological and developmental processes that these hormones influence.

Hormone Reception and Plant Development

A hormone receptor should be able to recognize subtle structural differences between small molecules in a highly specific manner. Binding of the hormone to the receptor should be noncovalent and reversible and result in an activated complex that induces the primary hormone response. Certainly, the plant hormone receptors reviewed here fulfill many of these criteria, and in the cases of ethylene, cytokinin, and BR perception, the classic modules of receptor kinases have been identified. Having the receptors in hand has allowed comparison to each other as well as to those in other systems. For example, ethylene interferes with its two-component receptor function, whereas the cytokinin two-component receptor acts through the archetypal mechanism of activation and phosphotransfer to a response regulator. The ethylene two-component kinase does not signal to a canonical response regulator, and its negative regulation of receptor function suggests the receptor has other functions in the absence of the hormone. In tomato, the BR receptor is not specific but is shared with another signaling pathway. Although the extracellular domain of BRI1 resembles the classic LRR–protein interaction motif, it has the capacity to bind an organic molecule through a novel embedded domain. Presumably the extracellular LRR originally served as a protein ligand receptor and the addition of the BR-binding ID domain arose secondarily.

These observations suggest that hormone receptors like those involving ethylene and BR signaling have additional functions that the hormones influence. Auxin and GA perception use a unique SCF-based proteolysis mechanism that takes advantage of F-box specificity for target proteins. However, in the case of GA-dependent targets, different hormones and environmental stresses such as salt can also influence target stability. The use of SCF-based proteolysis as a mechanism of signaling creates potential nodes of interactions between signaling molecules. For example, specific mutations that conferred auxin insensitivity and mapped to a Cullin subunit also altered jasmonate sensitivity by attenuating the assembly of the jasmonate specific F-box protein COI1 into the SCF complex.

These molecular mechanisms of hormone perception are now beginning to explain how these small enigmatic molecules can coordinate plant growth and development. Plant hormones have never had a clear location of synthesis and appear to cause a myriad of responses. For these reasons the role of hormones as "limiting factors" whose levels control a particular developmental pathway has never resonated well and has led to arguments that these small molecules actually function more as cofactors in balancing the different reactions and metabolic pathways comprising the overall network driving development. The non-Euclidian patterns of connectivity in a network not only buffer a system but also lead to the capacity to produce the same output via different strategies, which fits well with plant hormone phenomena. In this respect, that the loss of hormone perception does not generally result in overall lethality or even a reduction of cell types also supports the notion that hormones modulate rather than limit particular processes. Being perceived by proteins that have additional functions, or acting at central nodes of signalling partly supports a modulating versus control and command scenario for plant hormone function.

Whatever the case, it is clear there are differences between how plants and animals use organic-based hormones to organize tissues and organs. Unlike animals, plants are essentially an amalgamation of loosely dependent units that do not require highly centralized centers to coordinate growth and development. Because of this, different parts of a plant can experience different local environments and respond somewhat independently. Consequently, rather than having specific organs that make and respond to particular hormones, each independent unit produces and responds to many signals. By acting as factors that bind with high specificity to proteins with additional roles in development, plant hormones can add flexibility to signaling pathways that underlie the plasticity of development seen in higher plants.

Signaling Pathways

In plants, similar to that in animals, cells are in constant communication with one another. Plant cells communicate to coordinate their activities in response to the changing conditions of light, dark, and temperature that guide the plant's cycle of growth, flowering, and fruiting. Plant cells also communicate for coordination of their roots, stems, and leaves activities.

Elaborate signalling mechanisms are required in all complex multicellular organisms for both internal as well as external responses. Animals have two specialized systems for the purpose viz. nervous and endocrine system. But in plants, which lack motility, only hormonal system has developed which work as chemical messengers.

Plant signaling pathways differ in significant ways from those of animals in selection of trans membrane receptors, secondary messengers and dedicated mechanisms to environmental cues such as stress, drought, and light.

Ca^{2+}, inositol phospholipids, G proteins, cyclic nucleotides, protein kinases and protein phosphatases are common molecules present in majority of plants signaling pathways. G protein-coupled receptors though well characterized in animals and yeasts, plants do not have evidence of comparative abundance of the same. Plants have signaling pathways involving the hydrolysis of phosphatidylinositol -bis-phosphate [PtdIns P2] by plasmalemma-associated phospholipase C.

Many signal transduction pathways came into being when plants are challenged with various environmental stresses. It is still very much inconclusive as how to categorize these events. Specific networks for cold, drought and salt stress can be divided into three major categories: I) Osmotic/ oxidative stress signaling that makes use of MAPK modules, (II) Ca^{2+} -dependent signaling that lead to the activation of LEA-type genes (such as the DRE/CRT class of genes), and (III) Ca^{2+} -dependent SOS signaling that regulates ion homeostasis. Stress signaling knowledge is vital for development of rational breeding and transgenic strategies to improve stress tolerance varieties.

Cell differentiation in both shoot and root requires strict control of various cell populations which is regulated by specific signaling mechanisms. Coordinating expression of genes during plant development as well as during response to environmental cues is required elaborate signaling mechanisms. Such internal as well as external signaling agents bring about their effects by means of sequences of biochemical reactions, called signal transduction pathways, These pathways due to cascading effect greatly amplify the original signal and ultimately results in the activation or repression of genes.

Signaling cascades involve generation of second messengers inside the cell that greatly amplify the original signal. Among the most common second messengers in plant signaling pathways are 3′,5′-cyclic AMP (cAMP); 3′,5′-cyclic GMP (cGMP); nitric oxide (NO); cyclic ADP-ribose (cADPR); 1,2-diacylglycerol (DAG); inositol 1,4,5-trisphosphate (IP3); and Ca^{2+}.

Entire genomes of plants such as Arabidopsis and rice are now sequenced that leads to the possibility to search for unknown proteins by database mining. Such database searches of plant sequences have successfully identified family of receptor like protein kinases (RLK). Three types of RLKs have been identified in plants on the basis of extracellular domains.

Primarily plant development was known to be regulated by only five types of hormones: auxins, gibberellins, cytokinins, ethylene, and abscisic acid. However, with new set of evidences confirms presence of plant steroid hormones, the brassinosteroids. A variety of other signaling molecules having roles in resistance to pathogens and defense against herbivores are also been proved by experimentation, including jasmonic acid, salicylic acid, and the polypeptide systemin.

Phototropism is the phenomenon exclusive to plants resulted by differential growth due to bending towards light. Phototropism occurs by directional cell elongation, which is stimulated by the growth regulator auxin,

Thus plant signal transduction is diverse as animal system yet it lacks clear understanding of majority of the pathways. Plants are unique as compared to animals in the fact that their development and morphogenesis are plastic throughout their lives. They continuously monitor both biotic and abiotic parameters of their environment and these sensory perceptions shape their organs and bodies. Plants get their environmental cues via sensory perceptions which are critical for their survival. Especially light and gravity, two physical factors pervading the universe, are essential in this respect.

Despite having a relatively simple body organization, plants need sophisticated sets of coordinative processes of diverse functions such as root–shoot coordination, coordination amongst radial tissues, patterning in the totipotent plant cells to name few. Auxin is central in this respect that regulates patterning of cell differentiation, cell division and cell expansion.

Hormonal Signaling

Plant hormones control a diverse array of plant responses affecting growth and development, as well as defense against microorganisms and insects, and protection from abiotic. This complex process requires a communication system that can operate over relatively long distances among different plant organs as well as different organelles within a single cell. In such a system, cells of different tissues and organs are not only capable of detecting signals they receive from other parts of the plant, but also of responding and transmitting those signals in their own characteristic way. In higher organisms like plants, such diverse communication is performed by a group of chemical messengers called hormones.

A plant hormone is generally described as a naturally occurring organic compound that is active at very low concentrations (e.g.,<1 mM, often 1 uM). A hormone is often formed in certain parts of the plant and then translocated to other sites where it evokes specific biochemical, physiological, and/or morphological responses. These organic compounds promote, inhibit, or qualitatively modify plant growth and development in the tissues where they are produced as well as in distant tissues to which they are translocated. Therefore, the synthesis and action of plant hormones are not necessarily localized to a specific tissue, as with animal hormones, but occur in a wide range of tissues. In addition, plants respond to biotic and abiotic external stimuli such as pathogen and insect attack, drought, and salt stress using hormone signal transduction pathways that cause changes in the hormone metabolism and distribution within the plant.

The commonly recognized classes of plant hormones are auxin (IAA), gibberellin (GA), cytokinin (CK), abscisic acid (ABA), and ethylene (ACC). More recently recognized molecules involved in plant signaling include brassinosteroids (BR), jasmonic acid (JA), and salicylic acid (SA).

Hormone Signal Transduction Pathway

The induction of plant responses to any exogenous or endogenous stimuli requires a perception by the plant via different types of signal molecules collectively known as elicitors. Elicitors can be classified in 3 groups: (i) chemical signals such as hormones and phytotoxins, (ii) physical signals such as blue and red light, and (iii) biotic signals such as fungal elicitors. The chemical nature of these elicitors may vary from large molecules such as polypeptides, carbohydrates, glycoproteins, and fatty acids, to low molecular weight compounds such as hormones.

Another group of signal molecules that induce plant response to pathogens are those that can trigger defense responses at a distance from the inoculation site. Among the long-distance mobile signals, salicylic acid, jasmonic acid, and systemin are the most studied. Exogenous application of these compounds induces defense responses at a distance, and with SA there is an induction of protection against some challenge pathogens.

Signal transduction defines a specific information pathway within a cell that translates an intra- or extracellular signal into a specific cellular response. If the initial signal is a hormone, such as SA, GA, or ethylene, the first step in signaling involves the interaction of that hormone with a specific cellular recognition protein called a receptor. The initial phase of signal transduction requires high-affinity binding of the hormone to the receptor(s), which causes the receptor to undergo a conformational change that initiates a sequence of downstream events called signal transduction. After the signal is activated, the receptor may alter gene expression directly by acting as a transcription factor without transducing the activated signal to the pathway as in mammalian glucocorticoid receptors. Alternatively, the receptor may pass the signal to the nucleus through a series of intermediary steps acting as a molecular switch. In the pathway, the signaling components are generally modified by phosphorylation or by the activation of low molecular weight GTP-binding proteins. For instance, activation of nuclear factor-KB (NFKB) requires phosphorylation of a family of inhibitory proteins, IKBs via ubiquitination-dependent proteolysis, SCF E3RsasI$^{KBs/TrCP}$, which frees NF-KB to translocate to the nucleus where it regulates gene transcription in mammals. Similarly, SCF^{TIR} in auxin response suggests that similar phosphorylationbased signaling pathways might be involved. On the other hand, phosphorylation on a hydroxyl group of serine (Ser), threonine (Thr), or tyrosine (Tyr) residues is predominantly used in animals. In contrast to animal signal induced phosphorylation, a nitrogen atom of a histidine (His) residue and an acyl group of an aspartate (Asp) residue are predominantly used for phosphorylation in bacteria.

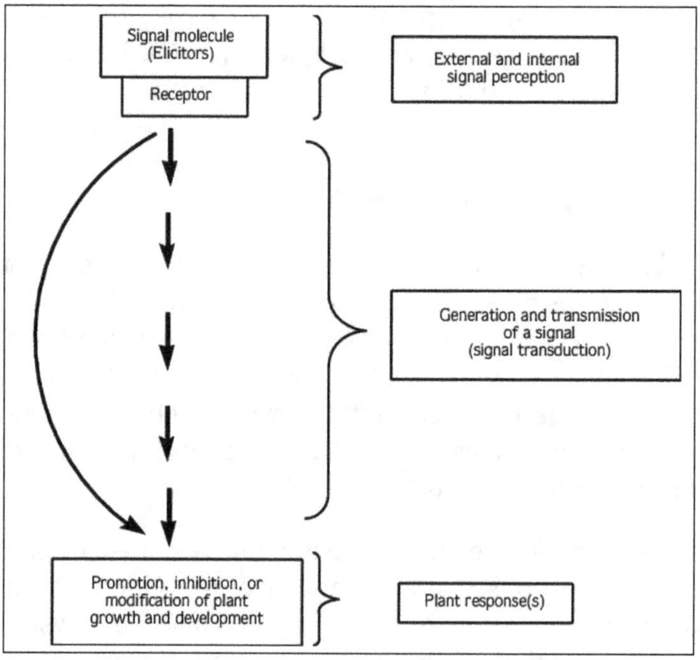

The phases of the hormone signaling pathway in plants, Three steps including signal perception, signal transduction, and plant response(s) are shown. An alternative pathway in which the receptor could alter gene expression directly by acting as a transcription factor without transducing the activated signal to the pathway is also shown.

Of the plant-specific signaling molecules including hormones, elicitors, and secondary metabolites, plants share some signaling agents with animals such as nitric oxide, reactive oxygen species, and other regulators function in both kingdoms. For instance, Glu, which was previously known as an animal signaling agent, is now regarded as a likely plant signaling compound, and genes encoding putative Glu receptor subunits have been identified in the Arabidopsis genome. This finding suggested that other low molecular weight compounds such as extracellular ATP (eATP) could be signaling agents in plants.

Since a signaling cascade can be a complex process, transduction pathways also require sensitivity and specificity that are coordinated and integrated with the related signaling components. Depending on the components of the pathway, the stimulation of the receptor must activate (positive) or inactivate (negative) relay components of the pathway through some type of cascading mechanism. In this case, the receptor acts as a molecular switch. These changes in signaling proteins not only permit a rapid response to the hormone signal but also allow recycling of components of the signaling system so that they can receive further signals. As a result, signal transduction not only modulates the enzyme activity in target cells, but also alters the rates of synthesis of existing proteins or triggers the synthesis of new ones.

Hormone Mutants Involved in Hormone Signaling Pathways

Plant hormone mutants can be classified into 2 main groups; (i) those that influence hormone levels by altering biosynthesis, generally termed biosynthesis mutants including (a) auxotrophs and (b) over accumulation mutants, and (ii) those that influence the response to hormones, generally termed response mutants including (a) insensitive and (b) hypersensitive mutants. Most auxotrophic mutants show a reduction in hormone level, and exogenous hormone application restores the mutant phenotype to its wild type. However, not all auxotrophs necessarily exhibit a reduction in the hormone biosynthesis. In some cases biosynthesis mutants may also overproduce hormones. On the other hand, response mutants appear to be insensitive to their own endogenous hormone levels or resistant to toxic or growth inhibiting levels of exogenous hormone. The main difference between a hormone response (insensitive or hypersensitive) mutant and a hormone biosynthesis (deficient) mutant is that the response mutant phenotype cannot be restored to the wild type by exogenous hormone application.

Another useful type of mutant in the investigation of complex hormone signaling is a secondary mutation that suppresses the effect of one of the mutations described above. Suppressors demonstrating their own phenotypes and partially suppressing an earlier gene mutation are useful not only for identifying new gene functions but also for identifying new mutations in previously characterized genes. Genes encoding components of a particular signaling pathway may have other functions that may be missed by direct screening but that can be identified genetically among suppressor mutations of signaling mutants. Recent studies have shown that this technique can identify novel genes functioning in the hormone signaling pathway in plants. For instance, a screen for suppressors of the auxin resistant mutant axr1 in Arabidopsis thaliana has identified a second site suppressor locus called SAR1 (Suppressor of Auxin Resistance 1). Genetic analysis of this mutant indicated that sar1 partially suppresses every aspect of axr1 and functions in the same or overlapping signaling pathway in auxin signalling.

To identify mutations in genes related to a specific hormone signaling pathway, the simplest and most used method is to assay a mutagenized plant population for an altered response to a specific hormone that is supplied exogenously. This should reveal a clear and reproducible phenotypic difference between wild type and mutant. However, in screens where seeds and seedlings are exposed to higher concentrations of hormone than those a plant experiences under normal growth conditions, mutations that confer insensitivity to such conditions may not always be specific to the hormone dependent pathway of interest. For instance, the iba1 (indole-3-butyric acid resistant 1) mutant of Nicotiana plumbaginifolia was recovered in a screen for resistance to a very low concentration of auxin, but was later found to be resistant to ABA and paclobutrazol, an inhibitor of gibberellic acid (GA) biosynthesis. In addition, not all hormone mutant genes determined in hormone screenings are necessarily directly involved in hormone signal transduction pathways. It is possible that mutations identified in a screen mark genes whose functions are necessary for a signaling event to occur, but which are not directly involved in the regulation of the signal transduction pathway. For instance, it has been suggested that early germination and the wilty phenotype of iba1 mutant are due to a change in the ABA/GA ratio; auxin may have a secondary effect on iba1 phenotype.

Mutants in hormone signaling genes can modulate (i) the level of receptors, (ii) the affinity of the receptor protein for the hormone, or (iii) the magnitude of the response. Insensitivity to a particular hormone may be attributed to a receptor that is uncoupled from the activating ligand, such as ETR1, to the effect of genes encoding biosynthetic enzymes that alter intracellular hormone levels, or to the effect of other genes whose actions in an unexpected activation of the hormone signal transduction chain such as in iba1 mutant. On the other hand, mutants that affect multiple hormones can shed light on the complex mechanisms through which hormone signaling is integrated in the plant.

In addition to the forward genetics approaches (i.e. beginning with a mutant phenotype and ending with the genetic sequence that causes the altered phenotype), the recent availability of the whole Arabidopsis genome sequence may provide an opportunity to use reverse genetics, such as insertional mutagenesis to resolve complex signaling pathways in plants. Reverse genetics begins with a mutant gene sequence and tries to identify the resulting change in the phenotype. Gene knockouts, or null mutations, provide a direct route to determining the function of a gene product in situ. New studies have shown that this approach can successfully identify novel mutants in plants). This approach usually involves the use of either transposable elements or T-DNA as a mutagen. The foreign DNA not only disrupts expression of the gene into which is inserted, but also acts as a marker for subsequent identification of the mutation because of its known sequence). An important aspect of this insertional mutation is that it permits the identification of genes that would have been missed in traditional mutagenesis screens because the success of traditional mutagenesis strictly depends on the selection methods applied to detect desired mutants). For instance, if a gene is functionally redundant, a reduction or loss of function of the gene may result in no obvious or only subtle phenotypic changes that cannot be identified in screens for mutant phenotypes but may be detected by expression pattern in enhancer-trap or gene-trap screens. In most traditional screens, since seeds and seedlings are exposed to higher concentrations of hormone than those plants experience, gene mutations that are homozygous lethal are usually missed, but can be maintained in the heterozygous plant populations with insertional mutagenesis.

Since jasmonate signaling has been one of the most extensively studied signaling pathways during the last decade, it was used an example to show how hormone mutants can be used to reveal complex hormone signaling in plants.

Ca^{2+} Based Signaling

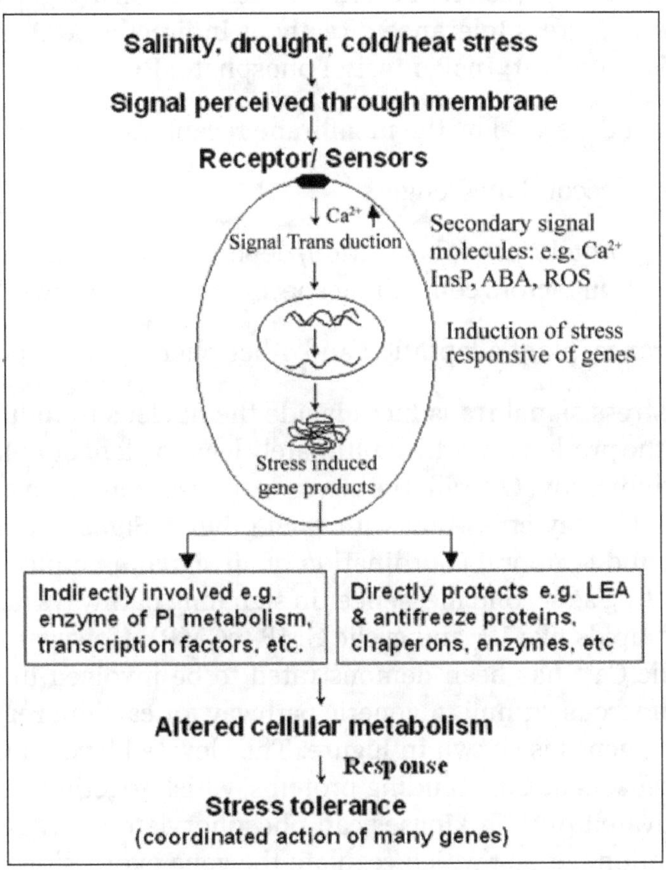

Calcium plays a fundamental role in plant's growth and development. Many extracellular signals and environmental cues including light, abiotic and biotic stress factors, elicit change in the cellular calcium levels, termed as calcium signatures.[1] Ca^{2+} has been well established as a second messenger and the concentration of Ca^{2+} is delicately balanced by the presence of 'Ca^{2+} stores' like vacuoles, endoplasmic reticulum, mitochondria and cell wall. Ca^{2+} is present in milimolar concentrations in the cell wall and vacuoles and is released whenever required by the cell. Researcher have demonstrated that the organelles surrounded by a double membrane (e.g. mitochondria, chloroplasts and nuclei) are also equipped to generate calcium signal on their own, which is delimited by a double membrane.

Ca^{2+} ion represents an important signaling molecule and a convergence point of many disparate signaling pathways. Facing the environmental challenge, plant cells reprogram their cellular set up by triggering a network of signaling events that start with stress perception at the membrane level of the cell and ends with a cellular response.

Generic pathway for plant response to stress, The extracellular stress signal is first perceived by the membrane receptors and then activate large and complex signaling cascade intracellularly including the generation of secondary signal molecules. The signal cascade results into the expression of multiple stress responsive genes, the products of which can provide the stress tolerance directly or indirectly. ABA, abscisic acid LEA, late embryogenesis abundant; InsP, inositol phosphate; ROS, reactive oxygen species.

- Perception of the signal by the membrane receptors.
- Generation of second messengers.
- A cascade of protein phosphorylation/dephosphorylation events that may target transcription factors controlling specific set of stress regulated genes.
- Stress tolerance, plant adaptation and other phenotypic responses.

The extracellular stress signal transduces inside the nucleus to induce multiple stress responsive genes, the products of which ultimately lead to plant adaptation to stress tolerance directly or indirectly. Overall, the stress response could be a coordinated action of many genes, which may cross-talk with each other's. Signal transduction requires the proper spatial and temporal coordination of all signaling molecules. In plant cell, many molecules acting as second messenger in signaling pathways have been reported; these include Ca^{2+}, lipids like IP_3 and cyclic GMP (cGMP). However no single messenger except cytosolic Ca^{2+} has been demonstrated to be involved in diverse pathways and respond to numerous stimuli. A generic pathway for calcium regulated expression of stress responsive genes is shown in Figure. The elevated level of Ca^{2+} can be recognized by calcium sensors or Ca^{2+}-binding proteins, which together can activate protein kinases. These activated protein kinases can phosphorylate many regulatory proteins including transcription factors, which regulate the gene expression level of regulatory proteins resulting in the altered metabolism followed by phenotypic response of stress tolerance. The response could also be growth inhibition or cell death, which will depend upon how many and what kinds of genes are getting upregulated or down-regulated in response to the elevated calcium.

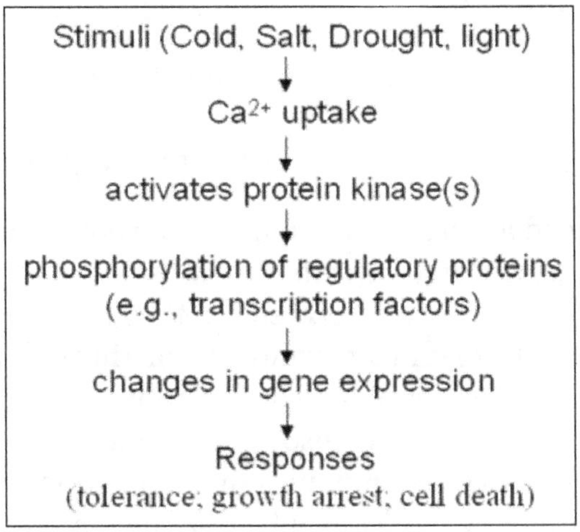

Figure shows A generic pathway for calcium regulated gene expression and the stress response. The calcium level increases in response to the stimuli. The increased level of Ca^{2+} is recognised by some Ca^{2+}-sensors or calcium-binding proteins, which can activate many calcium dependent protein kinases. These kinases regulate the function of many genes including stress responsive genes, resulted in the phenotypic response of stress tolerance.

Overall, in response to calcium several genes are reported to be upregulated. An analysis of transcriptome changes revealed 230 calcium-responsive genes, of which 162 were upregulated and 68 were downregulated. These include known early stress-responsive genes as well as genes of unknown function.[4] Recently, a blue light receptor phototropins, which regulate growth and development of plants, has also been shown to involved in calcium signaling in higher plants.[5] Ca^{2+} signaling pathway can also regulates a K^+ channel for low-K response in *Arabidopsis*. Calcium is also reported to be an essential component of the sucrose signaling pathway that leads to the induction of fructan synthesis.[6] Calcium signaling is also involved in the regulation of cell cycle progression in response to abiotic stress.

Role of calcium in plants:

- Calcium is an essential plant nutrient required for growth and development of plant, especially the root and shoot tip. The tips are meristmatic and cell division occurs by mitosis. Ca^{2+} helps in the formation of microtubules and microtubules in turn are essential for the anaphasic movement of chromosomes.
- Ca^{2+} is an important divalent cation and is required for structural roles in the cell wall and membranes where it exists as Ca^{2+} pectate. Ca^{2+} accumulates as calcium pectate in the cell wall and binds the cells together.

- It is also required as a counter-cation for inorganic and organic anions in the vacuole and as an intracellular messenger in the cytosol. Externally supplied Ca^{2+} reduces the toxic effects of NaCl and ameliorates stress.
- Ca^{2+} is required for pollen tube growth and elongation.

The Calcium Requirement of Plants and its Uptake from Soil

Calcium was first described as an essential macro nutrient element in plants more than a century ago. Ca^{2+} is taken up by roots from the soil solution and delivered to the shoot via the xylem. Ca^{2+} may traverse the root either through the cytoplasm of the cells, linked together by plasmodesmata (the symplast) or through the spaces between the cells (the apoplast). The cytosolic free Ca^{2+} concentration is maintained typically at 200 nM. However, the Ca^{2+} content of the cytosol is far higher than this because of the high affinity of Ca^{2+} to a range of Ca^{2+} binding proteins. Ca^{2+} mobility within the cell is very low as a consequence of its low free concentration and rapid chelation. It moves predominantly apoplastically rather than symplastically.

Plants vary markedly in their Ca^{2+} content and requirements. Ca^{2+} content can very between 0.1 and >5.0% dry wt. for different plants and organs. Soil Ca^{2+} also varies widely from <0.01% calcium in acid laterites to very high abundance in chalky soils.[9] Ecologists have classified plant species into calcifuges and calcicoles. Calcifuges are plants occurring on acidic soils with low Ca^{2+}. The term calciole is used to describe plants, which can grow on calcareous/Ca^{2+} rich soils. These soils are generally Ca^{2+} and base-rich soils. However, plants vary in their Ca^{2+} requirements and the ability to extract Ca^{2+} from complex soil environments. In particular, monocots require less Ca^{2+} for optimal growth than do the dicots. Ca^{2+} enters plant cells through Ca^{2+}-permeable ion channels in their plasma membranes. Ca^{2+} competes with other cations both for these sites and for the uptake from the soil. The presence of high levels of Ca^{2+} is known to ameliorate the effects of the uptake of toxic cations (Al^{3+} and Na^{+}) from the soil while the presence of high levels of other cations (K^{+}, Mg^{2+}) may reduce Ca^{2+} uptake. Calcium uptake and requirements for optimal growth are thus strongly dependent on the presence of other cations.

Calcium deficiency is rare in nature, but may occur on soils with low base saturation and/or high levels of acidic deposition. Calcium deficiency may occur because of competition by other cations, or because of low transpiration, where the xylem flow is inadequate to supply the calcium requirements of rapidly growing tissue. Calcium deficiency results in stunted root growth and altered leaf appearances. More severe symptoms generally result from the failure of cell membrane integrity and include bitter-pit in apple fruit, blossom end rot in tomato and tip burn in lettuce. Deficiency symptoms develop first in the regions of plant most distal to the region of Ca^{2+} uptake.

Ca^{2+} Transporters/Efflux Pumps in the Cellular Membranes

Plant cells, like animal cells, maintain a low regulated free Ca^{2+}concentration ranging from 30 nM to 400 nM in higher plants. Maintaining the low cytosolic Ca^{2+} concentration observed in plant cells requires active transport of Ca^{2+} from the cytosol. Active efflux pumping is a prerequisite for the restoration of low levels cytosolic calcium after the signaling event. The removal of Ca^{2+} from the cytosol against its electrochemical gradient to either the apoplast or to intracellular organelles requires energized 'active' transport. Ca^{2+}-ATPases and H^+/Ca^{2+} antiporters are the key proteins catalyzing this movement. By removing Ca^{2+} from the cytosol several important functions are performed by these enzymes.

- They maintain a low $[Ca^{2+}]_{cyt}$ in the resting or the unstimulated cell which is appropriate for the cytoplasmic metabolism.
- They restore $[Ca^{2+}]_{cyt}$ levels to the resting normal levels following a $[Ca^{2+}]_{cyt}$ perturbation, thereby influencing the magnitude kinetic and subcellular location of $[Ca^{2+}]_{cyt}$ signals.
- They replenish intracellular and extracellular Ca^{2+} stores for subsequent $[Ca^{2+}]_{cyt}$ signals and permit the generation of localized $[Ca^{2+}]_{cyt}$ oscillations through their interplay with Ca^{2+} channels.
- They provide Ca^{2+} in the ER for the secretory system to function.
- They remove some divalent cations, such as Ni^{2+}, Zn^{2+}, Mg^{2+} and Mn^{2+} from the cytosol to prevent mineral toxicity.

Suggested that the Ca^{2+}-ATPases, which have high affinity (Km = 1–10 μM) but low capacity for Ca^{2+} transport, are responsible for maintaining $[Ca^{2+}]_{cyt}$ homeostasis in the resting cells. Whereas the H^+/Ca^{2+}-antiporters, which have lower affinities (Km = 10–15 μM) but high capacities for Ca^{2+} transport, are likely to remove Ca^{2+} from the cytosol during $[Ca^{2+}]_{cyt}$ signals and thereby modulate $[Ca^{2+}]_{cyt}$ perturbations. This hypothesis is supported by the fact that H^+/Ca^{2+} antiporter, but not the vacuolar Ca^{2+}-ATPase, resets $[Ca^{2+}]_{cyt}$ in yeast following hypertonic shock.

Plant Ca^{2+}-ATPases belongs to two major families (a) The P-type ATPase II A family and (b) The P-type ATPase II B family.

- The P-type ATPase II A family: Nucleotide specificity of these pumps is broad (30–60% activity achieved with GTP and ITP). The pumps are inhibited by erythrosine B (IC_{50} ≤ 1 μM) and estimate Ca^{2+}affinity is in the range of 0.4–12 μM. The first family (The P-type ATPase II A family) lacks N-terminal auto regulatory domain. Four members of this family have been identified in the *Arabidopsis* genome.[16] They are likely to be present in the plasma membrane, tonoplast ER and the Golgi apparatus.
- The P-type ATPase II B family: The second family of plant Ca^{2+}-ATPases (the P-type ATPase IIB family) is characterized by an auto inhibitory N-terminal

domain that contains a binding site for Ca-CaM and in addition a serine-residue phosphorylation site. Their catalytic activity can be modulated by $[Ca^{2+}]_{cyt}$ either through activation upon binding CaM or by inhibition following phosphorylation by Ca^{2+}-dependent protein kinases (CDPK).[17] Since CaM binding sites are generally quite diverse, each type-II B Ca^{2+}-ATPase may have different affinity for CaM or may bind a different CaM isoform. Ten members of the type-II B, Ca^{2+}-ATPase family have been identified in the *Arabidopsis* genome (termed at *ACAS1, 2* and *4* and *ALACAs 7* to *13* by Axelsen and Palmgren, (2001).[16] These Ca^{2+}-ATPases reside on various cellular membranes including the plasma membrane (AtACA8), the tonoplast (AtACA4), and the plastid inner membrane (AtACA1). The relative molecular mass of type II B Ca^{2+}-ATPase pumps has been estimated to be between 115,000 Da and 135,000 Da.

The abundance of Ca^{2+}-ATPase isoforms suggest that individual isoforms are functionally distinct and may respond differentially to distinct cellular processes involving specific Ca^{2+} signals. They also imply a requirement for CaM-independent and CaM-dependent regulation of Ca^{2+}-ATPase activities in the modulation of $[Ca^{2t}]_{cyt}$ perturbations during cell signaling. The expression of many Ca^{2+}-ATPases is increased upon exposure to high salinity or high $[Ca^{2+}]_{cyt}$ and some Ca^{2+}-ATPase genes are expressed only under stress conditions.[19] This may reflect a role in maintaining $[Ca^{2+}]_{cyt}$ homeostasis or in reducing Na^+ influx to the cytosol in saline environments.

Ca^{2+}/H^+ Antiporters

In addition to a P-type pump, the presence of a low affinity Ca^{2+}/H^+ antiport mechanism at the plasma membrane level had been suggested. The first plant H^+/Ca^{2+} antiporter to be cloned and functionally expressed was *CAX1* (Calcium exchanger). The gene was identified by its ability to restore growth on a high Ca^{2+} medium to a yeast mutant defective in vacuolar Ca^{2+} transport.

The H^+/Ca^{2+}-antiporters present in the plasma membrane and tonoplast have been characterized biochemically. These have a lower affinity for Ca^{2+} than Ca^{2+}-ATPases and may also transport Mg^{2+}. The stoichiometry of the dominant H^+/Ca^{2+}-antiporter in the tonoplast is apparently $3H^+/1Ca^{2+}$. Eleven genes encoding putative H^+/Ca^{2+} antiporters (*AtCAX*) have been identified in the genome of *Arabidopsis thaliana*.[13] The transporters AtCAX1, AtCAX2 and AtCAX4 are located at the tonoplast.[13] The AtCAX1 antiporter exhibits both a high affinity and high specificity for Ca^{2+}. By contrast, the AtCAX2 transporter is a high-affinity, high capacity H^+/heavy metal cation antiporter. The *AtCAXs* have homologues in other plant species and their physiological roles have been investigated using transgenic plants.[13] Transgenic tobacco overexpressing AtCAX1 exhibits Ca^{2+}-deficiency disorders, which includes tip burn, metal-hypersensitivity and susceptibility to chilling that can be reversed by increasing Ca^{2+}supply. By raising Ca^{2+} supply the expression of *AtCAX1* and *AtCAX3* (but not *AtCAX2* or *AtCAX4*) genes was increased.

Ca^{2+}/H^+ antiporters, utilize the H^+ gradient generated by the tonoplast V-type H^+-pump and by a proton-pumping pyrophosphatase to sequester Ca^{2+} in the Vacuole.[21] In many plant cells the vacuole occupy more than 50% of the cell volume, and it is evident that trans tonoplast Ca^{2+} transport makes a very significant contribution to the regulation of cytosolic Ca^{2+} concentrations. Ca^{2+} concentrations within the vacuole range from 0.1 to 10 mM.

Ca^{2+}-ATPases are estimated to represent only <0.1% of the membrane protein and are thus 30–100 fold less abundant than H^+-ATPases in the PM (3%) and the endomembranes (5–10%). All Ca^{2+} pumps are inhibited by orthovanadate. H^+/Ca^{2+} antiporters are efflux transporter and are different from Ca^{2+}-ATPases in that they do not require ATP and they are not sensitive to vanadate.

In contrast Ca^{2+} influx to the cytosol is mediated via Ca^{2+} channels. The principal roles of Ca^{2+} permeable channels in the plasma membrane appear to be in all signaling. Ca^{2+} permeable channels have been found in all plant membranes. They have been classified on the basis of their voltage-dependence into depolarization activated (DACC), hyperpolarization-activated (HACC) and voltage independent cation channels (VICC).

The Ca^{2+} Signatures

The cytosolic Ca^{2+} in plant cells increases in response to various environmental challenges like abiotic and biotic stresses and developmental cues. This transient increase in $[Ca^{2+}]_{cyt}$ is considered critical for the production of a physiological response. Elevation of $[Ca^{2+}]_{cyt}$ is considered a universal response to stress. The perturbations in cytosolic calcium levels (termed as $[Ca^{2+}]_{cyt}$ "signature") elicited by each environmental challenge and developmental cue is unique and results in an appropriate physiological response to a particular stimulus. The uniqueness is manifested in the sub-cellular location and/or the kinetics of magnitude of the $[Ca^{2+}]_{cyt}$ perturbation.[23] An increase in $[Ca^{2+}]_{cyt}$ is effected by Ca^{2+} influx to the cytosol either from the apoplast, across the plasma membrane, or from the intracellular organelles. The Ca^{2+} influx is mediated by Ca^{2+} permeable ion channels, and their type, cellular localization and the abundance influences the spatial characteristics of $[Ca^{2+}]_{cyt}$ perturbations. Since the diffusion of Ca^{2+} within the cytoplasm is low, and the buffering of Ca^{2+} in the cytoplasm is high (0.1 to 1 mM) the opening of Ca^{2+} channel produces a local increase in $[Ca^{2+}]_{cyt}$ that dissipates rapidly after the channel has been closed. The subcellular localization of Ca^{2+} channel is therefore critical for the targeting of different cellular processes. Cytosolic calcium "waves" are produced within the cytoplasm by the successive recruitment of particular Ca^{2+} channels to coordinate cellular responses. It was suggested that a local elevation of $[Ca^{2+}]_{cyt}$ might generate soluble second messengers, such as IP_3 or cADPR, that diffuse through the cytoplasm to activate a relay of spatially separated Ca^{2+} channels.[24] This theory was supported in the plant cells responding to salt stress.

Researcher suggested that repetitive Ca^{2+} influx across the plasma membrane contributed in the $[Ca^{2+}]_{cyt}$ that occur following the application of ABA to guard cells. This transient increase was first close to the plasma membrane and subsequently adjacent to the vacuole. These waves are thought to reflect the sequential opening of hyperpolarization-activated Ca^{2+} channels at the plasma membrane and then second-messenger activated Ca^{2+} channels in the tonoplast.

In addition to these sub cellular waves, "waves" of cells with high $[Ca^{2+}]_{cyt}$ may also propagate through the plant tissue. This can be induced in root tissues by mechanical stimulation or saline shock, in cotyledons by cold shock and in leaves by chilling plant roots briefly. Electrical action potentials, osmotic perturbations or chemical signals may trigger these waves. Although an elevated $[Ca^{2+}]_{cyt}$ is necessary for signal transduction, a prolonged increase in $[Ca^{2+}]_{cyt}$ is lethal. Sustained high $[Ca^{2+}]_{cyt}$ is implicated in apoptosis, both during the normal development and in hypersensitive responses to pathogens.

To effect adaptive responses, the $[Ca^{2+}]_{cyt}$ perturbations must be either of low amplitude or transient. Transient increases in $[Ca^{2+}]_{cyt}$ can be single (spike), double (biphasic) or multiple (oscillations). The perturbations generated may differ in their cellular location, role and extent of propagation and on their amplitude during propagation. Calcium signatures may also be tissue specific. For example, within the root, the $[Ca^{2+}]_{cyt}$ perturbations induced by mechanical perturbation, salinity, osmotic stress, cold shock or slow cooling differ markedly between cell types.

Several abiotic challenges result in an immediate, transient increase in $[Ca^{2+}]_{cyt}$ that is restored to basal levels within minutes. Such changes include mechanical perturbations and rapid cooling for brief periods, termed as 'cold shock'. The duration, periodicity and amplitude of oscillations vary considerably, and their form is often dependent on the strength and combination of specific stimuli.

Calcium "Memory"

A diminished $[Ca^{2+}]_{cyt}$ elevation upon repetitive stimulation by the same environmental challenge or a developmental cue is a common observation. Many examples support this. The magnitude of the $[Ca^{2+}]_{cyt}$ perturbation elicited by the wind-induced motion becomes progressively smaller upon repeated stimulation and a refractory period of several minutes is required before a full response is observed again. A second exposure to an elicitor does not influence $[Ca^{2+}]_{cyt}$ for several hours after its initial application. For example, plant cells challenged with H_2O_2 fail to respond to H_2O_2 for several hours.There is also evidence that the $[Ca^{2+}]_{cyt}$ signatures elicited by one environmental challenge can be modified by prior exposure to a contrasting one. For example, the magnitude of the $[Ca^{2+}]_{cyt}$ perturbations in response to oxidative stress was reduced by prior exposure to hyperosmotic stress and vice-versa was also found to be true. These observations also imply a cross talk between the signaling cascades. The attenuated response of $[Ca^{2+}]_{cyt}$ after repeated stimulation by various elicitors forms a part of cellular memory and the cells are able to retain the previous information. This 'memory' is significant and helps the cells to respond better to a particular stress without disturbing the delicate balance of Ca^{2+} levels and maintaining cellular [Ca^{2+}] homeostasis. It is noteworthy that *Nicotiana plumbaginifolia*, which could not retain 'cold memory' during acclimation of the plant by pretreatment with nonfreezing cold temperature was cold-sensitive whereas *Arabidopsis* which could retain 'cold memory' was more resistant to cold stress.

Calcium Binding Proteins

Transient increase in the cytoplasmic Ca^{2+} in response to signals is sensed by an array of Ca^{2+} sensors. Ca^{2+} sensors are small proteins that bind Ca^{2+} and change their conformation in a Ca^{2+} dependent manner. Specificity in the signaling pathway is provided by the uniqueness in calcium signatures and also by plethora of Ca^{2+} sensors, which can decode the Ca^{2+} perturbations quite precisely. Once Ca^{2+} sensors decode the elevated $[Ca^{2+}]_{cyt}$, Ca^{2+}efflux into the cell exterior and/or the sequestration into cellular organelles such as vacuoles, ER and mitochondria restores its levels to resting state. The properties of binding proteins are described below.

Properties essential for Ca^{2+}-binding proteins (CaBPs) to function as intracellular-Ca^{2+}receptor:

- The Ca^{2+} receptor/sensor must have Ca^{2+}binding sites essentially unoccupied in the resting cell (free Ca^{2+} is 10^{-7}M) and occupied at levels, which are reached upon stimulation (~10^{-7}M – 10^{-6}M). This means that the affinity constants (Ka) for Ca^{2+} should be ~$10^{-6}M^{-1}$.

- The protein must show selectivity and preference for Ca^{2+} in the presence of other cations like Mg^{2+} and K^{+}.
- After binding to Ca^{2+}, a Ca^{2+} sensor must undergo a conformational change that either alters its interaction with other molecules or changes its activity if it is an enzyme. Formation of Ca^{2+} protein complex is reversible.
- The kinetics of interaction should be very fast so as to correspond with the short lived Ca^{2+} signature.

EF Hand, Structure and Function

Most of the Ca^{2+} sensors bind Ca^{2+} using a helix-loop-helix motif termed as the 'EF hand' or the elongation factor, which binds a single Ca^{2+} molecule with high affinity. The Ca^{2+} sensors utilize the side chain oxygen atoms of the EF hand motif for Ca^{2+} coordination.

- The EF hand motifs mostly exist in pairs, which helps in the stabilization of the protein structure. Exceptions to this norm include proteins such as, parvalbumin and *Plasmodium falciparum* membrane surface protein. Frequently, EF hand pairs interact through antiparallel β-sheets, which allow cooperation in Ca^{2+} binding.

- The EF hand is a highly conserved amino-acid motif consisting of a α helix E (residue 1–10), a loop (residue 10–21), which binds the Ca^{2+} ion and a second α-helix F (residue 19–29). The loop consists of residues with the pattern X (D), Y(D), Z (D), −Y (K), −X (D), −Z (E) which participate in binding Ca^{2+} and the intervening residues are represented by an asterisks (*). A representation of the EF hand and loop is shown in Figure respectively.

```
A En**nn**nx*y*ZG-YI-X**-Zn**nn**n
  123456789 012345 67890 123456789
  -------Helix E------ --------Loop-------- ------Helix ------

B X*Y*Z*-Y*-X**-Z
  DKDGDG KI D FE E
  ----------- Loop-----------
          (12 aa)
```

Figure shows (A) Helix-loop-helix pattern of an EF hand. The EF hand is a highly conserved 29 amino-acid motif consisting of a α helix E (residue 1–10), a loop (residue 10–21), which binds the Ca^{2+} ion and a second α-helix F (residue 19–29). (B) The canonical EF hand consensus sequence of the 12 amino acid Ca^{2+} binding loop. The loop consists of 12 residues with the pattern X (D), Y(D), Z (D), -Y (K), -X (D), -Z (E) which participate in binding Ca^{2+} and the intervening residues are represented by an asterisks (*).

- The Ca^{2+} ion is coordinated by an oxygen atom or by a bridging water molecule (−x) of the side chains of residues 10 (X), 12 (Y), 14 (Z) and 18 (−X). The ligand at vertex (−Y) is the carbonyl oxygen at residue 16. Position 21 (−Z) is usually Glutamate and is the sixth residue to coordinate Ca^{2+}.

- In most of the functional EF hand motifs, the first amino acid is Aspartate (Asp) and the twelfth is Glutamate (Glu), Glu contributes both its side chain oxygen atoms to the metal ion coordination. Since there are 7 oxygen ligands, the Ca^{2+} coordination exhibits pentagonal bipyramidal symmetry, whose axis is 10(X)–18(−X). Glycine (Gly), present at the apex of the loop is an invariant residue and allows the loop to take a sharp bend. Each loop is flanked on either side by 8–10 residues of alternating hydrophilic and hydrophobic character. This arrangement favors the formation of the outer and inner surface of an alpha helical cylinder respectively.

- Most of the EF hand proteins are characterized by the relatively high percentage of acidic residues (Troponin C, 29%, CaM, 25%, intestinal Ca^{2+} binding protein 23%, and parvalbumin, 18%). Several isoforms of an EF hand protein may exist in a single organism. The Ca^{2+} binding affinities of the EF hand protein vary substantially (Kd = 10^{-4}–10^{-9}M) and depend on the amino acid sequence of the protein, especially with regards to the 12 residue consensus loop that provide all the acids that directly ligate to Ca^{2+} ions. When Ca^{2+} binds to $[Ca^{2+}]_{cyt}$ sensors their structural and/or enzymatic properties change and their subsequent interactions with target proteins can alter enzymatic activities, cytoskeletal orientation, protein phosphorylation cascades and gene expression. It is believed that these changes result in stress tolerance or a developmental switch.

Important Calcium Sensors in Plants

The major family of Ca^{2+} sensors includes:

- Calmodulin (CaMs).
- Calmodulin like proteins.
- Calcium dependent protein kinases (CDPKs).
- Calcineurin B-like proteins.
- Calmodulin: CaM.

Calmodulin (CaM) is a small (17 KDa), highly conserved, acidic protein with two globular domains each containing two EF hands connected by a flexible α-helical linker.[39] CaM is found in the apoplast and in the cytosol, ER and the nucleus of plant cells. Within the cytosol, the estimated CaM concentration is 5 to 40 μM.[40] Role of CaM has been implicated in many physiological processes like light, gravity, mechanical stress, phytohormones, pathogens, osmotic stress, heat shock and chilling.

The structure of CaM was first solved by Babu et al., and revealed that all 4 EF hands are saturated by Ca^{2+} ions (4 Ca^{2+}). CaM appears to be regulatory protein and induces large changes in inter-helical angles as Ca^{2+} is bound. The affinity of CaM for Ca^{2+} is influenced by the presence of particular target proteins.[40] CaM can also regulate gene expression by binding to specific transcription factors.

- CaM-like proteins: Plants also possess CaM-like proteins, which differ from the CaM in containing more than 148 amino-acid residues and have between one to six EF hand motifs. They possess limited homology to CaM (75% identity with canonical CaM isoforms.[39] In *Arabidopsis*, they include: CaBP-22, TCH2 and TCH3, AtCP1, NADPH oxidases, and Ca^{2+} binding protein phosphatases such as ABI 1 and ABI 2. These proteins have been implicated in cellular responses to diverse environmental, developmental and pathological challenges.
- Ca^{2+}-dependent protein kinases: 5 types of Ca^{2+} regulated protein kinases have been reported in plants: (i) Ca^{2+} dependent and CaM independent protein

kinases (CDPKs); (ii) CDPK-related protein kinases (CRKs); (iii) CaM dependent protein kinases (CaMKs); (iv) Ca^{2+}/CaM-dependent protein kinases (CCaMK); (v) SOS3/CBL interacting protein kinases (SIPKs/CIPKs).

- Ca^{2+}-dependent and CaM-independent protein kinases CDPKs.

The CDPKs are ubiquitous in plants. There are at least 34 genes encoding CDPKs in the *Arabidopsis* genome[42] and similar numbers in other plant species. Apart from plants, *CDPKs* are also found in protozoans and algae. They generally have four EF hands at their C-terminus that bind Ca^{2+} and activate the serine/threonine kinase activity of the enzyme. These kinases require micromolar concentrations of Ca^{2+} for their activity and have no requirement of CaM or lipids. They have a unique structure as N-terminal protein kinase domain is fused with C-terminal auto-regulatory domain and a CaM like domain, which has Ca^{2+} binding EF hand or helix-loop-helix motif. The autoinhibitory domain of CDPKs is a 30 amino acid sequence, which acts as a pseudo-substrate.[43] The N-terminal domain of CDPKs is variable and provides specificity to different CDPK isoforms. These enzymes show several fold stimulation with Ca^{2+} and show autophosphorylation. The binding of Ca^{2+} to some of CDPKs, is modulated by lipids or phosphorylation.[42] Ca^{2+} binding to CDPK effects conformation of the kinase and relieve the inhibition caused by the autoinhibitory region.

CDPKs are implicated in pollen development, control of cell cycle, phytohormone signaling, light-regulated gene expression, gravitropism, thigmotropism, cold acclimation, salinity tolerance, drought tolerance and responses to pathogens.

CDPK-related Protein Kinases (CRKs)

CRKs are similar to CDPKs expect that the CaM-like region is poorly conserved with degenerate or truncated EF hands that may not be able to bind Ca^{2+}. There are at least seven CRKs in *Arabidopsis* genome, and orthologues of these are present in many plant species. However, the regulation and function of these kinases are not known.

- CaM kinases (CaMKs): Several CaMKs have been cloned from *Arabidopsis* and other plants.[46] Their kinase activity is stimulated by CaM dependent autophosphorylation and their catalytic activity is also modulated by CaM. They are highly expressed in rapidly growing cells and tissues of the root and flower.

Ca^{2+}/CaM-dependent Protein Kinases (CCaMKs)

These are a group of Ca^{2+}-dependent kinases, which in addition to Ca^{2+} also requires CaM for their activity. Thus CaM besides acting directly could also exert its effect by binding to protein kinases and modulating their activity. A Ca^{2+}/CaM-dependent protein kinase (CCaMK) was characterized from lily and other plant species.[47] Sequence analysis revealed the presence of an N-terminal catalytic domain, a centrally located CaM-binding domain and a C-terminal visinin-like domain containing only three EF hands. Biochemical studies of CCaMK established that Ca^{2+} and CaM stimulates CCaMK activity. In the absence of CaM, Ca^{2+} promotes autophosphorylation of CCaMK. The phosphorylated form of CCaMK possesses more kinase activity than the non-phosphorylated form.

SOS3/CBL Interacting Protein Kinases (SIPKs/CIPKs)

Calcineurin B-like proteins were found to interact specifically with a class of serine-threonine protein kinases known as CBL interacting protein kinases (CIPKs).[1, 2] Recently, a novel CIPK from pea has been reported and found to interact and phosphorylate the pea.

- Calcineurin B-like proteins: Calcineurin B-like proteins (CBLs) are relatively a new class of calcium sensors discovered in *Arabidopsis* originally, in search for the genes imparting salt tolerance and maintaining cellular ion homeostasis. Molecular analysis of the sos mutants opened a new chapter in relation to salt stress signalling that led to the discovery of a pathway that transduce a salt stress induced Ca^{2+} signal to reinstate cellular ion homeostasis. Currently 10 *CBL* and 25 CBL-interacting protein kinases (*CIPKs*) have been reported from *Arabidopsis*.[2] The *CBL-CIPK* network is also widely distributed among higher plants but except *Arabidopsis*, the complexity and characterization of this pathway remains largely unrevealed. As different plants vary in their genome complexity, phenotype and physiology therefore, specie specific function or functional diversification can be expected. The essential role imparted by *CBL-CIPK* genes in stress tolerance, necessitates their detailed characterization from higher plants. In fact, there has been no report on experimental characterization of CBL from any higher plant except *Arabidopsis*. *AtCBL3*, in particular has been largely overlooked even in *Arabidopsis*. the cloning and characterization of a novel CIPK and its interacting partner CBL from pea. Pea CIPK showed autophosphorylation and could phosphorylate pea CBL. Both pea CBL and CIPK were found to be coordinately upregulated in response to various stresses such as cold, salinity but not to dehydation stress.
- Ca^{2+}-binding proteins without EF hands.

There are several proteins that bind Ca^{2+} but do not contain EF hand motifs. These include the phospholipase D (PLD), annexins, calreticulin and Pistil-expressed Ca^{2+} binding protein (PCP).

Phospholipase D

The activity of PLD, which cleaves membrane phospholipids into a soluble head group and PA, is regulated by $[Ca^{2+}]_{cyt}$ through a Ca^{2+}/phospholipids binding-site termed as the 'C2 domain'. PLD activity is implicated in cellular responses to ethylene and ABA, α amylase synthesis in aleurone cells, stomatal closure, pathogen responses, leaf senescence and drought tolerance.[51] Plants posses several PLD isoforms that differ in their affinity for Ca^{2+} and their modulation by phosphoinositides, free fatty acids and lysolipids. These biochemical modulators of PLD activity are the substrates or products of PLC, which generates IP_3, DAG, phospholipase A_2 and DAG-kinase, both of which are regulated by CaM. It is suggested that $[Ca^{2+}]_{cyt}$ signaling cascades might coordinate the activities of these diverse enzymes to effect specific responses to the environmental stimuli.

Annexins

These are a family of proteins in plants and animals that bind phospholipids in a Ca^{2+} dependent manner and contain four to eight repeats of approximately 70 amino acids. Although exact function of annexins is not known, plant annexins are implicated in secretory processes and some have ATPase, peroxidase activities.

Calreticulin

It is a Ca^{2+} sequestering protein in the ER and functions as a molecular chaperone. The function of calreticulin is also implicated in Ca^{2+} homeostasis.

Pistil Expressed Ca^{2+}-Binding Protein (PCP)

A 19 kDa novel Ca^{2+} binding protein (PCP) expressed in anthers and pistil was isolated. PCP is high capacity (binds 20 mol of Ca^{2+} per mol of PCP), low affinity Ca^{2+} binding protein. Role of PCP has been implicated in pollen-pistil interactions and/or pollen development.

G Protein Mediated Signaling

Heterotrimeric G proteins ae key components in regulation signal of transduction in animals as well as plants. However, compared to animals, plant G proteins are structurally and functionally unique. A major distinguishing factor is that plants lack a specific G protein coupled receptor which in animals is involved in perceiving and transferring the external signals. In the recent times, numerous studies have identified a wide range of functions of G protein in plants mainly using over-expression and mutant analysis.

Overall, plant G protein studies clearly exhibits its possibilities to identify novel targets in improving agricultural yield, disease resistance, and reducing growth period, to name a few.

Plant and Animal G Proteins: Similarities and Differences

G protein is a membrane associated heterotrimer that consists of Gα, Gβ and Gγ subunits. Plants generally have a small repertoire of G proteins. For instance, Arabidopsis has one canonical Gα subunit, G PROTEIN α SUBUNIT 1 (GPA1), three non-canonical Gα like subunit, EXTRA LARGE G PROTEIN 1, -2 and -3 (XLG 1, XLG 2 and

XLG 3), one Gβ, ARABIDOPSIS G PROTEIN β SUBUNIT 1 (AGB 1) and three Gγ, ARABIDOPSIS G PROTEIN γ SUBUNIT 1, -2 and -3 (AGG1, AGG 2, and AGG 3) have been identified. In the contrary, in humans, more than 23 Gα, 5 Gβ and 12 Gγ proteins have been identified so far.

In animals, G protein is associated to a 7 transmembrane G protein coupled receptor (GPCR). The activation of G protein occurs when an extracellular ligand interacts with the GPCR. This interaction leads to the conversion of GDP associated with the Gα subunit to GTP followed by the dissociation of Gα-GTP and Gβγ dimer. This results in further downstream signalling processes. Regulation of G protein function is mediated via the intrinsic GTPase activity of the Gα subunit, that leads to the protein inactivation.

In contrast to human G proteins, plant G proteins are not associated with GPCR. Furthermore, the plant Gα subunits have a low GTPase activity, hence are mostly bound to GTP, leading to a constitutively active state Therefore, the regulation of plant G protein requires GTPase accelerating proteins (GAPs) for the protein inactivation. The most important GAP in Arabidopsis, ARABIDOPSIS REGULATOR OF G PROTEIN SIGNALLING 1 (At RGS1) have a 7TM domain in their N terminal and a cytosolic RGS box domain in the C terminal. In the inactive state, G protein is bound to At RGS1, which activates the GTPase leading to the GTP hydrolysis.

Studies have also shown that RGS activity is modulated via plant specific Leucine rich repeat receptor-like kinases (LRR-RLKs). RLKs composes of an extracellular receptor domain, a single pass transmembrane domain and a serine threonine kinase. The phosphorylation of RGS mediated via RLKs is an essential event in RGS activity. This phosphorylation leads to a change in the global confirmation of RGS-Gα interaction, thus disabling its intrinsic GTPase activity. Further, the modified RGS detaches from the G protein and is removed via endocytosis.

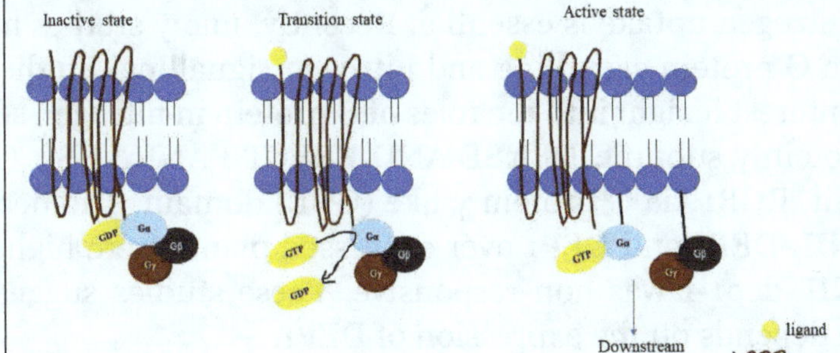

Protein signalling in animals.

In the inactive state, the Gα subunit is associated with GDP and is seen as heterotrimer. With the protein action by ligand binding on GPCR, the GDP is converted to GTP leading to downstream signalling events.

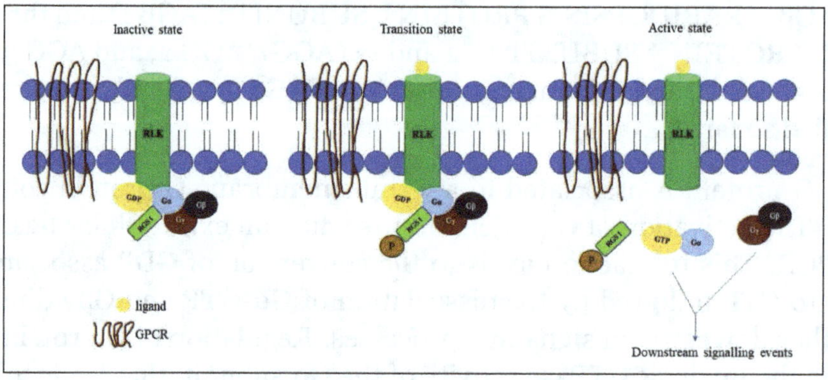

Snapshot of plant G protein signalling.

In the inactive stage, Gα subunit is bound by RGS1 and GDP. Unlike animals, G protein is associated with RLK in plants. Upon activation by a ligand, RGS1 is phosphorylated and is degraded and the disintegrated G protein activates further downstream.

In addition to the Gα subunit, many plants have a non-canonical Gα, extra-large G protein (XLG). XLGs interacts with Gβγ dimer to form a heterotrimer. However, unlike Gα they do not interact with RGS but displays a higher affinity to Gβγ. Moreover, they are phosphorylated directly, via kinases leading to the dissociation of the trimer and further downstream signalling.

Regulation of Nitrogen in Plants by G Proteins

With the advent of green revolution, we have seen a massive rise in the use of nitrogen fertilisers. Although the increased use of nitrogen fertilisers reported a dramatic increase in agricultural productivity, the soil quality has since been highly compromised. Nitrogen is one of the most important plant macronutrients, and is a constituent of majority of plant components. Hence, sustainable ways of nitrogen fertilisation and improved plant nitrogen uptake is essential. Recently, many studies have suggested a crosstalk between G protein signalling and nitrogen signalling, Studies in rice G proteins suggest an interest insight into the roles of G protein in nitrogen sensing in plants. In rice, the G protein γ subunit, DENSE AND ERECT PANICLES 1 (DEP1) interacts with the β subunit, RGB1 via G protein γ like (GGL) domain. The near isogenic lines (NIL) of DEP1, NIL-DEP1 and DEP1 over expressed plants were highly responsive to nitrogen while NIL-dep1-1 was non-responsive. These studies suggest that nitrogen sensitivity in rice depends on the expression of DEP1.

However, although dep1 were non-responsive to nitrogen, they showed an increased yield and root to shoot ratio. Another study that used CRISPR Cas 9 technique to generate DEP mutants, also showed an increase in number of grains. These studies suggest that, DEP1 functions depends on the interactions with other 'favourable genes', that could differ between cultivars.

G proteins are also known to have significant roles in the development of nodules in legumes further confirming the importance of G proteins in nitrogen assimilation. In soybean, nod factors like the End40 and cytokine oxidase, act as ligands for the Nod Factor Receptor 1 (NFR1), thus activating the NFR1 kinase domain. This leads to the phosphorylation of RGS and dephosphorylation of the Gα subunit and formation of the heterotrimer. Active Gα is a negative regulator of nodulation, it also interacts with NFG1 independent of RGS while the Gβγ dimer promotes nodulation in soybean.

Interestingly, studies have shown that nitrogen fixation in non-legumes also occurs via Rhizobium infection. However, the mechanism of infection is still unclear. Considering the role of G proteins in Rhizobium mediated nitrogen fixation in legumes, the possibility of a similar mechanism in nonlegumes cannot be excluded. Comparing the importance of DEP1 in nitrogen assimilation in rice, it is possible that DEP 1 is involved in recognising rhizobial elicitors in rice and requires further study.

Furthermore, it is also reported that with increase in the expression of RGS, rates of nodulation also increases. However, the study does now overlook at any other downstream elements that could be involved in the regulation of the signalling cascade. Also, the possibility of other upstream regulatory genes that controls the dephosphorylation of Gα is still unexplored.

Studies on nodulation in Lotus, also found significant roles of G protein signalling in the process. It has been known for long, that the dephosphorylation of Gα subunit involves a physical interaction with the G couples receptor 1(GCR1). In Lotus, the downregulation of LjGCR1 was found to have a significant effect on the genes involved in nodulation, NODULE INCEPTION GENE, LjNIN. This clearly suggest that LjNIN is a downstream element in the G protein signalling cascade. Further, LjNIN encodes for transcriptional factors essential for nodulation, further confirming its role as a downstream element. A negative feedback that might be exhibited by the downstream proteins to regulate the rate of dephosphorylation, as seen in animals, cannot be exempted. The role of G protein in plant-microbe interaction further indicates its possible involvement in plant immunity, about which is discussed further.

G Protein Regulates Plant Immunity

Many studies have shown significant cross talks between plant immune signalling and G proteins. In Arabidopsis, pathogen recognition receptor (PRR), FLS2 is an extensively studied aspect of innate immunity. Studies by Liang showed that XLG2, AGB1 andAGG1/2 is closely involved in FLS2 mediated plant innate immunity. According to their study, PRR mediated immune signalling in Arabidopsis involves a pre activation and post activation stage.

The pre activation stage is responsible for maintaining an optimum immune activation. This involves, a strong interaction of XLG2 with FLS2 and BIK1. During this phase,

XLG2 functions together with AGB1 and AGG1/2 thus forming a heterotrimer resulting in a proteasome mediated degradation of BIK1 leading to a stable immune activation. During a bacterial attack, FLS2 identifies the bacterial flagellin epitope, flg. This leads to the dissociation of XLG2 from the heterotrimer and is phosphorylates by BIK1. This phosphorylation leads to enhanced ROS production mediated by RbohD, a NADPH oxidase, thus leading to immune response.

Interestingly, a recent study by Liang clearly illustrates the mechanism of BIK1 mediated XLG2 phosphorylation. The study suggests that, during the pre-activation stage, a GTPase activating protein, RGS1 maintains the heterotrimer in its inactive stage. However, during the postactivation state, interaction of FLS2 with flg stimulates the phosphorylation of RGS1 leading to its dissociation from FLS2- G protein complex, which leads to immune reaction. This study brings in a novel insight of de-repression followed by release of G protein as a mechanism of immune response.

Researcher also hypothesised that the phosphorylated XLG2 mediates the ROS production. However, recently studies suggest further complicated machinery involving MAP4K SIK1in mediating ROS burst. SIK1 was found to interact with BIK1 and phosphorylates RBOHD leading to ROS production. This also provides insights into new genetic targets to develop plant resistance.

Reports also suggest that with respect to changes in the pathogen elicitors, the signalling pathways vary drastically. For instance, upon activation by bacterial PrpL(PA14 protease IV)/ArgC proteases, Mitogen Activated Protein Kinase, MAPK is activated. MAPK is an important element in downstream regulation in G protein signalling. Further analysis showed that MAPK is linked to the immune pathway via RACK1 (RECEPTOR FOR ACTIVATED C KINASE 1), a Gβ homologue. Contrast to this, flg is sensed by the FLS2-BAK complex and initiates the downstream processes.

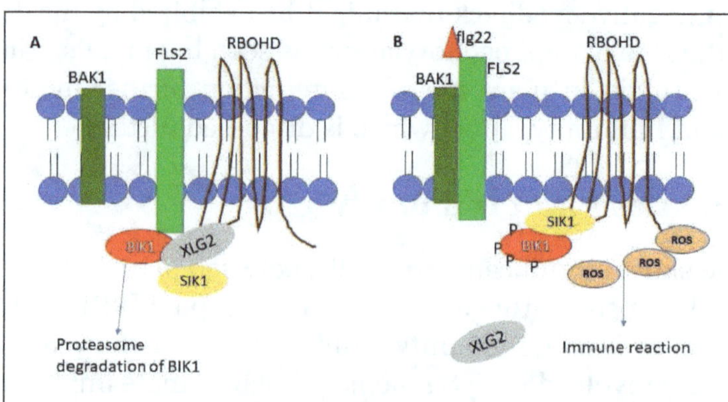

G protein mediated immune response.

(A) During the absence of a bacterial elicitor flg, XLG2 occurs as a heterotrimer. It is bound to BIK1 and is in a state of inactivation. Further, BIK1 undergoes proteasomal degradation. (B) When FLS2 senses flg, the resulting signal transduction leads to phosphorylation of BIK1 via SIK1. Also, XLG2 detached from the heterotrimer. This further mediates production of ROS by modulating RBOHD leading to immune reactions.

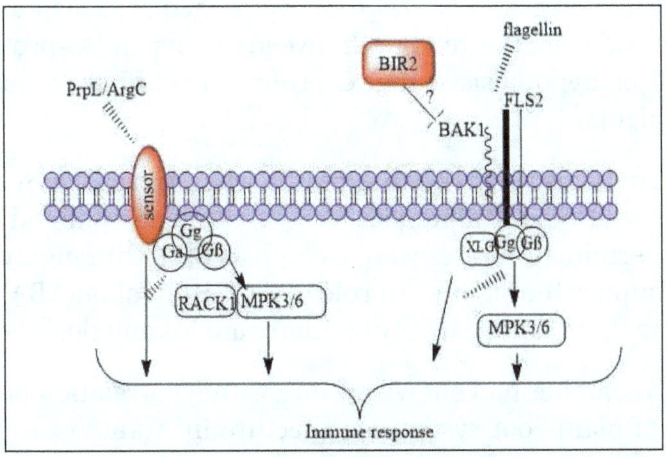

Differences in G protein associated immune responses with differences in pathogen elicitors.

Interaction of PrpL/ArgC elicitors is perceived via the membrane receptors leading to G protein association. This leads to activation of MPK3/6-RACK1 and further immune response activation (left). With the interaction of flagellin, MPK3/6 mediated response occurs via G protein activation leading to further downstream immune reactions. The possibility of BIR2 interacting with BAK1 that could also lead to G protein activation needs to be studied (right). (Ga stands for Gα and Gg stands for Gγ subunit).

Regulation of Plant Hormones by G Protein

Many studies have previously demonstrated that G proteins have an important role in regulation of plant hormone responses. In previous studies, mutated G proteins showed impaired growth and development and higher sensitivity to auxins, gibberellins, jasmonic acids, abscisic acid (ABA) and brassinosteroids (BR).

In plants, Gγ subunit is much diverse and has three distinct types. A recent study by researcher, showed the functional characterisation of type B Gγ in tomato, (SlGGB1). Fluorescence studies showed that SlGGB1 is localised in nucleus, cytoplasm and plasma membrane and had a strong interaction with Gβ subunit. However, Gβ subunits are generally localised in the plasma membrane, suggesting that SlGGB1 and Gβ interaction possibly occurs in the plasma membrane. This also implies that the interaction triggers a series of signalling reactions that leads to transcriptional activation in the nucleus.

Mutant studies showed that Gγmutant tomatoes had an increased number of lateral roots. This suggests an increase in the rate of cell elongation or increased lateral root primordia (LRP), which could be attributed to changes in plant growth hormone, auxin. Further studies showed increased auxin concentration in the mutants. These results implicate that auxin signalling is negatively regulated by SlGGB1. However, the mechanism of regulation is yet to be studied.

Recent study that looked at the role of iron nanoparticles (FeNp) in plants, showed a similar increase in the lateral root formation upon treatment with varying concentrations of FeNp. the increase in lateral roots were mediated via auxin transportation via SUX1 gene. Considering the relation between auxins and G protein in lateral root formation, it could be hypothesised that G proteins could have a possible role in iron transportation in plants.

Interestingly, the study also showed that plants when treated with ABA, SlGGB1 was over expressed. In wild type Arabidopsis, ABA treatment inhibited seed germination whereas in slggb1 germination rates were as high as 82% with highly supressed lateral root growth. This implies the significant role of SlGGB1 in plant ABA response and recognition, suggesting that GB1 negatively regulates auxins and positively regulated ABA.

Another study by researcher had shown an interesting correlation between the growth and maintenance of plant root system architecture in Arabidopsis, via sugar sensing by G proteins. In agb1, Gβ mutant, glucose levels were high compared to the wild type. In addition, there was also an increase in lateral roots in the mutants, suggesting that, G protein negatively regulate lateral root formation. Further study showed high levels of an auxin transport facilitator, PIN2, in the mutants. This suggests that G proteins mediate glucose transport and maintains root system architecture through via auxin signalling.

Brassinosteroids (BRs) are an important group of plant growth promoting hormones. They are involved in cell division and elongation and functions via G protein signalling. Recently a study by researcher showed an interesting crosslink between G protein signalling and BR signalling pathway. The study demonstrated that, in Arabidopsis, AGB1 interacts with BRI1-EMS SUPRESSOR 1 (BES1) via altering their phosphorylation state. BES1 is active in its dephosphorylated state, where it transports to the nucleus along with BZR1 to bind with BR1 promoter and initiate BR responses. Together, these studies demonstrate the complexity of G protein signalling and its possible cross talks with other important plant signalling pathways.

Stress Signal Transduction

Plants experience many different kinds of abiotic stresses including higher concentration of salt (salinity), extremes of temperature (low temperature, i.e., cold [chilling or freezing], higher temperature [heat] and water shortage (drought or dehydration). These stresses are the principal cause of reducing the yield of crops significantly. Overall, stress tolerance is a complex phenomenon because plants may go through multiple stresses at the same time during their development. Among abiotic stresses, the high salinity stress is the most severe environmental stress, which impairs crop production on at least 20% of irrigated land worldwide. In addition, increased salinity of arable land is expected to have devastating global effects, resulting in up to 50% land loss by the middle of 21st century.

Plants perceive and respond adaptively to abiotic stress imposed by salt, cold, drought and wounding and the adaptive process is controlled mainly by the phytohormone, abscisic acid (ABA), which acts as an endogenous messenger in the regulation of the plant's water status. ABA is generated as a signal during a plant's life cycle to control seed germination and developmental processes. The action of ABA can target specifically guard cells for induction of stomatal closure but may also signal systemically for adjustment towards severe water shortage. Since various stresses induce ABA synthesis, it is now considered as a plant stress hormone. Various transcription factors are known to regulate the ABA-responsive gene expression. Stress responsive genes can be expressed either through an ABA-dependent or ABA-independent pathway.

Generic Pathway for Plant Response to Stress

Plants can respond to stresses as individual cells and synergistically as a whole organism. The extracellular stress signal is first perceived at membrane level by the membrane receptors, ion channel, receptor-like kinase (RLK) or histidine kinase (HK) etc and then activate large and complex signaling cascade intracellularly including the generation of secondary signal molecules such as Ca^{2+}, inositol phosphates (InsP), reactive oxygen species (ROS) and ABA. The stress signal then transduces inside the nucleus to induce multiple stress responsive genes, the products of which ultimately lead to plant adaptation to stress tolerance directly or indirectly.1 Overall, the stress response could be a coordinted action of many genes, which may cross-talk with each other's. The stress-induced gene products are also involved in the generation of regulatory molecules like ABA, salicylic acid and ethylene, which can initiate the second round of signaling. Small molecules like ABA play important role in this process.

ABA and Abiotic Stress Signaling

ABA is an important phytohormone and plays a critical role in response to various stress signals. The application of ABA to plant mimics the effect of a stress condition. As many abiotic stresses ultimately results in desiccation of the cell and osmotic imbalance, there is an overlap in the expression pattern of stress genes after cold, drought, high salt or ABA application. This suggests that various stress signals and ABA share common elements in the signaling pathway and these common elements cross-talk with each other, to maintain cellular homeostasis. ABA also play important roles in many other physiological processes such as seed dormancy and delays its germination, development of seeds, promotion of stomatal closure, embryo morphogenesis, synthesis of storage proteins and lipids, leaf senescence and also defense against pathogens. ABA levels are induced in response to various stress signals. ABA actually helps the seeds to surpass the stress conditions and geminate only when the conditions are conducive for seed germination and growth. ABA also prevents the precocious germination of premature embryos. Stomatal closure under drought conditions prevents the intracellular water loss and thus ABA is aptly called as a stress hormone.

Main function of ABA seems to be the regulation of plant water balance and osmotic stress tolerance. Several ABA deficient mutants namely aba1, aba2 and aba3 have been reported for Arabidopsis. ABA deficient mutants for tobacco, tomato and maize have also been reported Without any stress treatment the growth of these mutants is comparable to wild type plants. Under drought stress, ABA deficient mutants readily wilt and die if the stress persists. Under salt stress also ABA deficient mutants show poor growth. In addition, ABA is required for freezing tolerance, which also involves the induction of dehydration tolerance genes.

Studies suggest that osmotic stress imposed by high salt or drought is transmitted through at least two pathways; one is ABA-dependent and the other ABA-independent. Cold exerts its effects on gene expression largely through an ABA-independent pathway. ABA induced expression often relies on the presence of cis-acting element called ABRE element (ABA-responsive element). Genetic analysis indicates that there is no clear line of demarcation between ABA-dependent and ABA-independent pathways and the components involved may often cross talk or even converge in the signaling pathway. Calcium, which serves as a second messenger for various stresses, represents a strong candidate, which can mediate such cross talks. Several studies have demonstrated that ABA, drought, cold and high salt result in rapid increase in calcium levels in plant cells.

The transcript accumulation of RD29A gene is reported to be regulated in both ABA-dependent and ABA-independent manner.Proline accumulation in plants can be mediated by both ABA-dependent and ABA-independent signaling pathways. The role of calcium in ABA-dependent induction of P5CS gene during salinity stress has been reported by. It is known that the expression of RD29A, RD22, COR15A, COR47 and P5CS genes was reduced in the los5 mutant. The signaling mechanism behind the activation of these genes are not well known, but the transcriptional activation of few stress induced genes represented by RD29A is known to some extent. It is also suggested that phospholipase D (PLD) along with ABA and calcium act as a negative regulator of proline biosynthesis in Arabidopsis. The salinity stress-induced upregulation of transcript of pea DNA helicase 45 (PDH45) followed ABA-dependent pathway while calcineurin B-like protein (CBL) and CBL-interacting protein kinase (CIPK) from pea followed the ABA-independent pathway. Overall, the ABA-dependent pathways are involved essentially in osmotic stress gene expression. Researcher have proposed that the activation of inactive ABA pools by polymerized AtBG1 (a β-glucosidase, hydrolyzes glucose-conjugated) could be a mechanism by which plants rapidly adjust ABA levels and respond to changing environmental cues.

ABA Biosynthesis

ABA level is known to induce under stress condition, which is mainly due to the induction of genes for enzymes responsible for ABA biosynthesis. Several ABA biosynthesis genes have been cloned which includes Zeathanxin epoxidase (Known as ABA1 in Arabidopsis), 9-cis-epoxycarotenoid dioxygenase (NCED), ABA aldehyde oxidase and

ABA3 also known as LOS ABA is synthesized from β-carotene through several enzymatic steps. The abiotic stress-induced activation of many ABA biosynthetic genes such as zeaxanthin oxidase (ZEP), 9-cis-epoxycarotenoid dioxygenase (NCED), ABA-aldehyde oxidase (AAO) and molybdenum cofactor sulphurase (MCSU) appeared to be regulated through calcium-dependent phosphorylation pathway. The accumulation of ABA can also feedback stimulate the expression of ABA biosynthetic genes through calcium-signaling pathway and can also activate the ABA catabolic enzymes to degrade the ABA. The mechanisms by which the abiotic stress upregulate ABA biosynthesis genes are not well understood. Researcher suggested that metabolic changes that alter hydrogen peroxide levels could also affect both ABA accumulation and ABA sensitivity.

ABA and Transcription Factors in Stress Tolerance

Transcriptional regulatory network of cis-acting elements and transcription factors involved in ABA and abiotic stress responsive gene expression is depicted in figure. The promoters of the stress-induced genes contain cis-regulatory elements such as DRE/CRT (A/GCCGAC), ABRE (PyACGTGGC), MYC recognition sequence (MYCRS; CANNTG) and MYB recognition sequence (MYBRS; C/TAACNA/G), which are regulated by various upstream transcriptional factors. The ABA-dependent stress signaling activates basic leucine zipper transcription factors called AREB, which binds to ABRE element to induce the stress responsive gene (RD29B). In Arabidopsis, it is reported that two ABRE motofs are involved in the regulation of ABA-responsive expression of the RD29B gene, which encodes a LEA-like (late embryogenesis abundant) protein. Transcription factors like DREB2A and DREB2B trans-activate the DRE cis-element of osmotic stress genes and thereby are involved in maintaining the osmotic equilibrium of the cell. Some genes like RD22 lack the typical CRT/DRE elements in their promoter suggesting their regulation by some other mechanism. RD22 gene encodes a protein having a homology to an unidentified seed protein. The drought-inducible expression of DREB1D is regulated by ABA-dependent pathways, indicating that DREB1D protein may function in the slow response to drought that depends upon the accumulation of ABA. The MYC/MYB transcription factors, RD22BP1 and AtMYB2, could bind MYCRS and MYBRS elements, respectively and help in activation of RD22 gene.

These MYC and MYB proteins are known to synthesize only after endogenous levels of ABA accumulate, hence suggesting that their role is in a late stage of the stress responses. The Clp protease regulatory subunit encoding gene, ERD1 (early responsive to dehydration), responds to dehydration and salinity stress before the accumulation of ABA, which suggested that an ABA-independent pathway also exists in the dehydration stress response of Arabidopsis. Overexpression of both ZF-HD and NAC proteins activate the expression of ERD1 gene in normal growth conditions (non-stressed) in the transgenic Arabidopsis plants. Overall, these transcription factors may also cross-talk with each other for their maximal response to stress tolerance. Researcher reported that overexpression of a transcription factor regulating ABA-responsive gene expression confered multiple stress tolerance.

The ABA level goes down during seed imbibition, which allow embryos to germinate and develop into seedlings, while ABA level remains high during abiotic stress conditions, which can arrest the growth and development. Several transcription factors, including abscisic acid-insensitive, ABI3 and ABI5, are known to control this developmental checkpoint. Recently, researcher have shown that in germinating Arabidopsis thaliana seeds, ABA induces the accumulation of microRNA 159 (miR159) in an ABI3-dependent fashion and miRNA159 mediates cleavage of MYB101 and MYB33 transcripts in vitro and in vivo. Here, they have shown that, the two MYB transcription factors function as positive regulators of ABA responses, as null mutants of myb33 and myb101 show hyposensitivity to the hormone. These results suggested that ABA-induced accumulation of miR159 controls transcript levels of two MYB factors during Arabidopsis seed germination. This is a homeostatic mechanism to direct MYB33 and MYB101 transcript degradation to desensitize hormone signaling during seedling stress responses.

Secondary Messengers

Multicellular organisms cannot survive without efficient communication among cells, tissues and organs. Regulation and coordination of metabolism, growth, morphogenesis depends of chemical signals from one part of the plant to another. German botanist Julius von Sachs is credited for originating this idea in the late nineteenth century.

Typically the end result of any signaling pathway is alteration in gene expression via regulating the transcription factors. Signal transduction pathways often involve the generation of second messengers, transient secondary signals inside the cell that greatly amplify the original signal. Binding to a receptor initiates a signal transduction pathway, often involving the generation of second messengers, such as cyclic nucleotides, inositol triphosphate, and calcium, Such pathways normally lead to changes in gene expression. Chemical signals specifically bind protein receptors found either on the plasma membrane or in the cell's cytoplasm.

Second messengers are molecules that relay signals received at receptors on the cell surface such as the arrival of protein hormones, growth factors, etc. to target molecules in the cytosol and/or nucleus. In addition to their job as relay molecules, second messengers serve to greatly amplify the strength of the signal. Binding of a ligand to a single receptor at the cell surface may end up causing massive changes in the biochemical activities within the cell.

There are 3 major classes of second messengers:

- Cyclic nucleotides (e.g., cAMP and cGMP).
- Inositol trisphosphate (IP3) and diacylglycerol (DAG).

- Calcium ions (Ca^{2+}).

There are three basic types of secondary messenger molecules:

- Hydrophobic molecules: Water-insoluble molecules such as diacylglycerol, and phosphatidylinositols, which are membrane-associated and diffuse from the plasma membrane into the intermembrane space where they can reach and regulate membrane-associated effector proteins.
- Hydrophilic molecules: Water-soluble molecules, such as cAMP, cGMP, IP3, and Ca^{2+}, that are located within the cytosol.
- Gases: Nitric oxide (NO), carbon monoxide (CO) and hydrogen sulfide (H2S) which can diffuse both through cytosol and across cellular membranes.

These intracellular messengers have some properties in common:

- They can be synthesized/released and broken down again in specific reactions by enzymes or ion channels.
- Some (such as Ca^{2+}) can be stored in special organelles and quickly released when needed.
- Their production/release and destruction can be localized, enabling the cell to limit space and time of signal activity.

3′,5′-Cyclic AMP (cAMP)

Adenosine 3':5'-monophosphate, commonly known as cAMP, is a key second messenger in living organisms ranging from Dictyostelium to Homo sapiens. cAMP was first described as a regulator of glycogen breakdown in the liver and is now known to be a second messenger for a wide variety of cellular responses in animals.

3′,5′-Cyclic GMP (cGMP)

It has been shown that the second messenger cyclic guanosine 3,5-monophosphate (cGMP) is an important signalling component with multiple biological functions in plants similar to the situation in animals. In plants, there is some biochemical evidence supporting the existence of plant cGMP-responsive kinases, and the Arabidopsis genome contains sequences that encode gene products with both a cyclic nucleotide binding domain and a protein kinase. However, specific cGMP targets in plants are largely unknown and in particular there is little molecular evidence available of bona-fide cGMP-dependent kinases.

Nitric Oxide (NO)

Nitric Oxide has multifunctional roles in plants: stomatal movement, host-pathogen interactions, hormonal signaling during growth/development and adaptation to abiotic/

biotic stress. NO in guard cells plays a very important role as a secondary messenger during stomatal closure induced by effectors.

Cyclic ADP-ribose (cADPR)

There is wide range of physiological functions are signalled and regulated by mobilization of intracellular Ca^{2+} stores. The generally accepted view is that cells possess multiple types of stores that can be mobilized by specific messenger molecules. Two unrelated nucleotides with Ca^{2-} mobilizing activity was identified viz. cyclic ADP-ribose (cADPR) and nicotinic acid adenine dinucleotide phosphate (NAADP). The action of cADPR on the channel is temperature-dependent, and proposed to be as a temperature sensing mechanism in cells.

1,2-Diacylglycerol (DAG)

Diacylglycerol (DAG) has unique functions as a basic component of membranes, an intermediate in lipid metabolism and a key element in lipid-mediated signaling.

Inositol Trisphosphate (IP3)

Inositol trisphosphate or inositol 1,4,5-trisphosphate (also commonly known as triphosphoinositol; abbreviated InsP3 or Ins3P or IP3), together with diacylglycerol (DAG), is a secondary messenger molecule used in signal transduction and lipid signaling in biological cells. While DAG stays inside the membrane, IP3 is soluble and diffuses through the cell. It is made by hydrolysis of phosphatidylinositol 4,5- bisphosphate (PIP2), a phospholipid that is located in the plasma membrane, by phospholipase C (PLC).

Increases in the intracellular Ca^{2+} concentrations are often a result of IP3 activation. When a ligand binds to a G protein-coupled receptor (GPCR) that is coupled to a Gq heterotrimeric G protein, the α-subunit of Gq can bind to and induce activity in the PLC isozyme PLC-β, which results in the cleavage of PIP2 into IP3 and DAG.

If a receptor tyrosine kinase (RTK) is involved in activating the pathway, the isozyme PLC-γ has tyrosine residues that can become phosphorylated upon activation of an RTK, and this will activate PLC-γ and allow it to cleave PIP2 into DAG and IP3. This occurs in cells that are capable of responding to growth factors such as insulin, because the growth factors are the ligands responsible for activating the RTK.

IP3 (also abbreviated Ins3P) is a soluble molecule and is capable of diffusing through the cytoplasm to the ER, or the sarcoplasmic reticulum (SR) in the case of muscle cells, once it has been produced by the action of PLC. Once at the ER, IP3 is able to bind to the Ins3P receptor (Ins3PR) on a ligand-gated Ca^{2+} channel that is found on the surface of the ER. The binding of IP3 (the ligand in this case) to InsP3R triggers the opening of the Ca^{2+} channel, and thus release of Ca^{2+} into the cytoplasm. In heart muscle cells this increase in Ca^{2+} activates the ryanodine receptor-operated channel on the SR, results in further increases in Ca^{2+} through a process known as calcium-induced calcium release. IP3 may also activate Ca^{2+} channels on the cell membrane indirectly, by increasing the intracellular Ca^{2+} concentration.

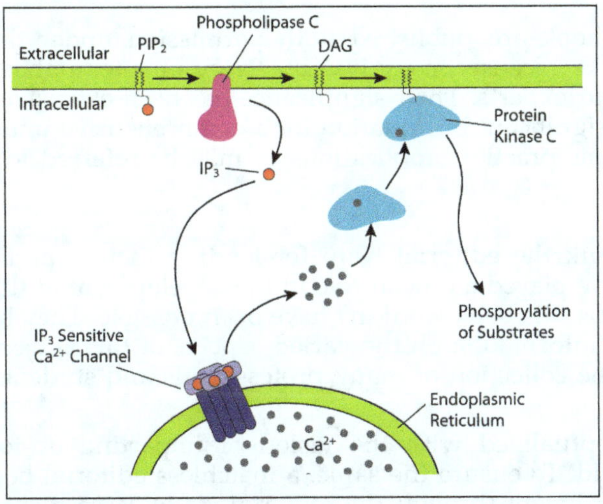

Ca^{++}

Evolution has adopted positively charged calcium as the primary signaling element of cells. Ca^{2+} is one of the principal second messengers for functioning as a central node in the overall "signaling web" and plays an important role in providing stress tolerance to plants.

Analysis of [Ca^{2+}] in dynamics has demonstrated its signaling role in plant cells in response to a wide array of environmental cues. In general, the stress leads to increased cytosolic Ca^{2+}, which initiates the stress signal transduction pathways for the stress tolerance. Ca^{2+} is the most tightly regulated ion within all membrane-bound organisms and binds to several proteins to effect changes in localization, association, and function. It is now clear that Ca^{2+} signaling affects almost every aspect of the cellular metabolism of living organisms.

In fact, an indirect increase in [Ca^{2+}] cyt in plant cells is mediated by the increase in intracellular concentrations of other second messengers such as cyclic GMP, cyclic AMP, and IP3, in response to stimuli. The elevation of Ca $^{2+}$ concentration in cytoplasm is found to be a key event in the plant cell for transduction of various signals to a biological effect.

Permissions

All chapters in this book are published with permission under the Creative Commons Attribution Share Alike License or equivalent. Every chapter published in this book has been scrutinized by our experts. Their significance has been extensively debated. The topics covered herein carry significant information for a comprehensive understanding. They may even be implemented as practical applications or may be referred to as a beginning point for further studies.

We would like to thank the editorial team for lending their expertise to make the book truly unique. They have played a crucial role in the development of this book. Without their invaluable contributions this book wouldn't have been possible. They have made vital efforts to compile up to date information on the varied aspects of this subject to make this book a valuable addition to the collection of many professionals and students.

This book was conceptualized with the vision of imparting up-to-date and integrated information in this field. To ensure the same, a matchless editorial board was set up. Every individual on the board went through rigorous rounds of assessment to prove their worth. After which they invested a large part of their time researching and compiling the most relevant data for our readers.

The editorial board has been involved in producing this book since its inception. They have spent rigorous hours researching and exploring the diverse topics which have resulted in the successful publishing of this book. They have passed on their knowledge of decades through this book. To expedite this challenging task, the publisher supported the team at every step. A small team of assistant editors was also appointed to further simplify the editing procedure and attain best results for the readers.

Apart from the editorial board, the designing team has also invested a significant amount of their time in understanding the subject and creating the most relevant covers. They scrutinized every image to scout for the most suitable representation of the subject and create an appropriate cover for the book.

The publishing team has been an ardent support to the editorial, designing and production team. Their endless efforts to recruit the best for this project, has resulted in the accomplishment of this book. They are a veteran in the field of academics and their pool of knowledge is as vast as their experience in printing. Their expertise and guidance has proved useful at every step. Their uncompromising quality standards have made this book an exceptional effort. Their encouragement from time to time has been an inspiration for everyone.

The publisher and the editorial board hope that this book will prove to be a valuable piece of knowledge for students, practitioners and scholars across the globe.

Index

P

S

T

V

X